HOLT CALIFORNIA
Earth Science

Study Guide A
with Directed Reading Worksheets

HOLT, RINEHART AND WINSTON
A Harcourt Education Company
Orlando • **Austin** • New York • San Diego • London

TO THE STUDENT

Do you need to review the concepts in the text? If so, this booklet will help you. The *Study Guide* is an important tool to help you organize what you have learned from the chapter so that you can succeed in your studies. The booklet contains a Directed Reading worksheet and a Vocabulary and Section Summary worksheet for each section of the chapter.

Use these worksheets in the following ways:

- as a reading guide to identify and study the main concepts of each chapter before or after you read the text
- as a place to record and review the main concepts and definitions from the text
- as a reference to determine which topics you have learned well and which topics you may need to study further

ISBN-10: 0-03-099393-8
ISBN-13: 978-0-03099-393-0
3 4 5 6 7 018 11 10 09 08 07

Contents

Directed Reading A

Section: Thinking Like a Scientist (pp. 8–15)
SCIENTIFIC HABITS OF MIND

Write the letter of the correct answer in the space provided.

_____ **1.** What are some habits of mind that scientists share?
  a. They are critical, open, and unethical.
  b. They are close-minded, dull, and uninspired.
  c. They are dishonest, intellectual, and unimaginative.
  d. They are curious, skeptical, and ethical.

Curiosity

_____ **2.** What did scientist Jane Goodall's curiosity lead to?
  a. She studied chimpanzees for more than 30 years.
  b. She studied humans for more than 50 years.
  c. She studied horses for more than 30 years.
  d. She studied birds for more than 30 years.

Skepticism

_____ **3.** What is skepticism?
  a. the ability to keep an open mind
  b. the ability to be curious
  c. the ability to be creative
  d. the practice of questioning accepted ideas

_____ **4.** What was Rachel Carson skeptical about?
  a. claims made about pesticides
  b. claims made about chimpanzees
  c. claims made about earthquakes
  d. claims made about cold fusion

Openness to New Ideas

_____ **5.** What does keeping an open mind mean?
  a. discouraging curiosity
  b. considering new ideas
  c. discrediting scientists
  d. disproving scientific facts

Imagination and Creativity

Match the correct description with the correct term. Write the letter in the space provided.

_____ **6.** guides scientists during their research

_____ **7.** helps scientists think about the world in new ways

_____ **8.** prevents scientists from lying about experiments

_____ **9.** ensures the honesty of published work

_____ **10.** ensures that people choose to take part in research, based on their understanding of risks

a. creativity

b. honesty

c. peer review

d. ethics

e. informed consent

WHAT DOES A SCIENTIST LOOK LIKE?

Write the letter of the correct answer in the space provided.

_____ **11.** What is scientist Mae Jemison doing now?
    a. studying the ozone layer
    b. developing treatments for AIDS
    c. improving lives in West Africa
    d. changing our thinking about the universe

Scientific Literacy

_____ **12.** What is scientific literacy?
    a. understanding methods of scientific inquiry
    b. learning to distrust information
    c. thinking illogically
    d. learning to read

Logic and Analysis

_____ **13.** What can becoming scientifically literate give you?
    a. a way to make money
    b. skills to be popular
    c. a closed mind
    d. skills for daily life

_____ **14.** Which is NOT something that science can teach you?
    a. to make careful observations
    b. to think logically about information
    c. to be a better-informed consumer
    d. to be more emotional

Critical Thinking and Science

_____ **15.** What is the key to critical thinking?
   a. dismissing information without evaluation
   b. disbelieving all information
   c. evaluating the information you find
   d. taking your best guess

SCIENCE IN OUR WORLD

_____ **16.** Which of the following best describes science?
   a. It happens only in the laboratory.
   b. It happens only in the classroom.
   c. It is not a process.
   d. It affects everyone.

Scientists as Citizens

_____ **17.** What does the ozone layer do?
   a. damages CFCs
   b. protects living things from harmful radiation
   c. protects people from CFCs
   d. helps increase refrigeration

_____ **18.** What happened as a result of the work of Mario Molina?
   a. CFCs were banned.
   b. Pollution research was banned.
   c. New chemicals were discovered.
   d. CFCs were allowed in Europe.

FROM THE CLASSROOM TO THE WORLD
Special Science Programs

_____ **19.** What is something teenagers can do at special science programs in California?
   a. Design, build, and "fly" ROVs.
   b. Study owl nests.
   c. Discover new chemicals.
   d. Travel to outer space.

Classroom Collaboration

_____ **20.** What is the JASON Project?
   a. a field trip to Florida
   b. a virtual two-week research trip
   c. an ocean research project
   d. a science fair

Skills Worksheet

Directed Reading A

Section: Scientific Methods in Earth Science (pp. 16–23)
LEARNING ABOUT THE NATURAL WORLD

Write the letter of the correct answer in the space provided.

_____ 1. What is the beginning of a process scientists use to learn about the natural world?
   a. collecting data
   b. communicating results
   c. asking questions
   d. studying dinosaurs

WHAT ARE SCIENTIFIC METHODS?

_____ 2. Which of the following do scientists use to answer questions and solve problems?
   a. the natural world
   b. dinosaurs
   c. scientific methods
   d. hypothesis

_____ 3. Which of the following describes scientific methods?
   a. a series of steps to solve problems
   b. a series of problems to be solved
   c. a series of answers to questions
   d. a series of hypotheses

_____ 4. Which statement about the steps of scientific methods is correct?
   a. Scientists always use them in order.
   b. Scientists never repeat steps.
   c. Scientists must use all the steps.
   d. Scientists use the steps in ways that work best.

ASKING A QUESTION

_____ 5. What does asking a question help a scientist do?
   a. analyze results
   b. focus the purpose of an investigation
   c. answer a question
   d. communicate results

_____ **6.** What question might David Gillette have asked when he examined some dinosaur bones?
- **a.** Who found the bones?
- **b.** What kind of dinosaur did they come from?
- **c.** When were the bones found?
- **d.** Where were the bones found?

FORMING A HYPOTHESIS

_____ **7.** What is a hypothesis?
- **a.** a question
- **b.** an experiment
- **c.** a variable
- **d.** a possible answer to a question

_____ **8.** Which of the following is NOT true of a hypothesis?
- **a.** It must be able to be tested.
- **b.** It is sometimes called an *educated guess*.
- **c.** It is an experiment.
- **d.** It is a scientist's best answer to a question.

Making Predictions

_____ **9.** What is a statement in an if-then form called?
- **a.** a prediction
- **b.** data
- **c.** a variable
- **d.** a question

TESTING THE HYPOTHESIS

_____ **10.** How do scientists test hypotheses?
- **a.** by using their imagination
- **b.** by gathering data
- **c.** by asking questions
- **d.** by guessing

_____ **11.** What are pieces of information gathered through observations or experimentation called?
- **a.** investigation
- **b.** prediction
- **c.** data
- **d.** hypothesis

| Directed Reading A *continued*

Testing with Experiments

Use the terms from the following list to complete the sentences below.

variable   controlled experiment

12. Only one factor at a time is tested in a(n) _____.

13. Another name for a factor in an experiment is a(n) _____.

Testing Without Experiments

Write the letter of the correct answer in the space provided.

_____ **14.** When scientists cannot use a controlled experiment to test something, what do they depend on?
   a. variables
   b. predictions
   c. observations
   d. hypotheses

_____ **15.** What does it mean when large amounts of data support a hypothesis?
   a. The investigation should be given up.
   b. The hypothesis is probably correct.
   c. The experiment was done wrong.
   d. The hypothesis is not valid.

ANALYZING THE RESULTS

_____ **16.** What does analyzing results help scientists do?
   a. collect data from experiments
   b. form questions and hypotheses
   c. form explanations based on evidence
   d. do controlled experiments

_____ **17.** During which step do scientists make tables and graphs?
   a. analyzing results
   b. asking questions
   c. forming a hypothesis
   d. making a prediction

DRAWING CONCLUSIONS

_____ **18.** Which of the following statements is true when a hypothesis is not supported?
   a. The question was not valid.
   b. The observations were wrong.
   c. The results are still valuable.
   d. The investigation will end.

_____ **19.** What was David Gillette's conclusion about the bones he studied?
   a. The dinosaur did not exist.
   b. The bones were from an unknown dinosaur.
   c. The hypothesis was wrong.
   d. The bones were not dinosaur bones.

COMMUNICATING RESULTS

_____ **20.** Why do scientists communicate their results?
   a. to stop experiments from being repeated
   b. to stop the use of scientific methods
   c. to share what they have learned
   d. to stop scientists from changing hypotheses

_____ **21.** How do openness and repeating experiments help a scientist?
   a. They maintain a scientist's believability.
   b. They prove the scientist is right.
   c. They help the scientist earn money.
   d. They show that the scientist is wrong.

Is the Case Closed?

_____ **22.** If the discovery of a new skeleton does not support David Gillette's hypothesis, what may happen?
   a. The dinosaur bones will be thrown away.
   b. The investigation will not continue.
   c. A new hypothesis may be formed.
   d. A new dinosaur will be discovered.

Skills Worksheet

Directed Reading A

Section: Safety in Science (pp. 24–29)
THE IMPORTANCE OF SAFETY RULES
Write the letter of the correct answer in the space provided.

_____ **1.** What do safety rules help prevent?
    a. experiments and conclusions
    b. science and learning
    c. accidents and injury
    d. injury and diseases

Preventing Accidents

_____ **2.** What is the most important safety rule when you do science?
    a. Follow directions.
    b. Don't ask questions.
    c. Don't do experiments.
    d. Wear goggles.

Preventing Injury

_____ **3.** If there is an accident, how can you avoid or reduce injuries?
    a. Run out of the classroom.
    b. Follow safety rules after an accident.
    c. Don't tell anyone.
    d. Scream and jump up and down.

ELEMENTS OF SAFETY
Safety Symbols

_____ **4.** What do safety symbols tell you?
    a. how to bake cookies
    b. how to prevent injury or accidents
    c. how to do experiments
    d. how to leave the classroom

_____ **5.** What should you learn about each safety symbol?
    a. what it warns you about
    b. what color it is
    c. where to find it
    d. how to draw it

| Directed Reading A *continued*

READING AND FOLLOWING DIRECTIONS

_____ **6.** What should you always do before a science activity?
    a. Wash your hands.
    b. Talk to your best friend about it.
    c. Read all instructions very carefully.
    d. Take off your safety equipment.

_____ **7.** If you don't understand directions in the science lab, who should you ask to explain them?
    a. the school principal
    b. your parents
    c. your best friend
    d. your science teacher

Neatness

_____ **8.** Why should you arrange your materials neatly during an experiment?
    a. to make your work area look nice
    b. to please your teacher
    c. so you find them easily
    d. so you can go home early

Using Proper Safety Equipment

_____ **9.** What should you do if you need to handle hot objects?
    a. Use your apron.
    b. Ask your friend to handle them.
    c. Call the teacher.
    d. Wear heat-resistant gloves.

Proper Clean-Up Procedures

_____ **10.** What should you do with extra chemicals after an activity?
    a. Do what the teacher tells you to do with them.
    b. Ask your friend what to do with them.
    c. Take them home.
    d. Wash them down the drain.

| Directed Reading A *continued*

Match the correct description with the correct term. Write the letter in the space provided.

_____ **11.** clearing books off your work area

_____ **12.** wiping your work area with wet paper towels

_____ **13.** wearing goggles and an apron

_____ **14.** knowing what the symbol of a small animal means

_____ **15.** doing what the instructions and your teacher say

a. recognizing safety symbols

b. following directions

c. practicing neatness

d. using the right safety equipment

e. cleaning up properly

PROPER ACCIDENT PROCEDURES

Steps to Follow After an Accident

Write the letter of the correct answer in the space provided.

_____ **16.** Which of the following is NOT a step to follow after an accident?
   a. Stay calm and assess what happened.
   b. Leave the area.
   c. Inform the teacher or call for help.
   d. Assist the teacher with cleanup or first aid.

_____ **17.** What should you always do if an accident happens?
   a. Run out of the classroom.
   b. Keep on doing the activity.
   c. Bring food to the injured person.
   d. Tell your teacher.

Caring for Injuries

Use the terms from the following list to complete the sentences below.

first-aid kit   first aid

18. Emergency medical care for someone who has been hurt or who is sick is

called _____.

19. You should be familiar with how things in a(n) _____,

such as bandages, are used.

Skills Worksheet

Vocabulary and Section Summary A

Thinking Like a Scientist

VOCABULARY

In your own words, write a definition of the following terms in the space provided.

1. skepticism

_____

_____

_____

2. scientific literacy

_____

_____

_____

SECTION SUMMARY

Read the following section summary.

- Scientists are curious, creative, skeptical, and open to new ideas.
- It is important for scientists to be honest and ethical in their treatment of humans and other living things.
- People from diverse backgrounds have made many contributions to the advancement of science.
- Increasing scientific literacy and developing critical-thinking skills are goals of science education.
- Scientists always evaluate the credibility of information that they receive.
- Scientists can have public roles in society. In addition to explaining scientific concepts to the media, scientists work to improve the quality of people's lives.
- There are many opportunities to participate in science programs in your community.

Skills Worksheet

Vocabulary and Section Summary A

Scientific Methods in Earth Science
VOCABULARY

In your own words, write a definition of the following terms in the space provided.

1. scientific methods

_____

_____

2. hypothesis

_____

_____

3. data

_____

_____

4. controlled experiment

_____

_____

SECTION SUMMARY

Read the following section summary.

- Scientific methods are the ways in which scientists follow steps to answer questions and solve problems.

- The steps used in scientific methods are to ask a question, form a hypothesis, test the hypothesis, analyze the results, draw conclusions, and communicate results.

- A controlled experiment tests only one factor at a time so that scientists can determine the effects of changes to just that one factor.

- Accurate record keeping, openness, and replication of results are essential to maintaining a scientist's credibility.

- When similar investigations give different results, the scientific challenge is to verify by further study whether the differences are significant.

Vocabulary and Section Summary A

Safety in Science

VOCABULARY

In your own words, write a definition of the following term in the space provided.

1. first aid

_____

_____

_____

SECTION SUMMARY

Read the following section summary.

- Following safety rules helps prevent accidents and helps prevent injury when accidents happen.
- Five elements of safety are recognizing safety symbols, following directions, being neat, using safety equipment, and using proper cleanup procedures.
- Animals used in scientific research require special care.
- When an accident happens, assess what happened, secure the area, report the accident, and help care for injuries or help clean up.
- First aid is emergency medical care for someone who has been hurt.

Skills Worksheet

Directed Reading A

Section: Tools and Measurement (pp. 44–49)

Write the letter of the correct answer in the space provided.

_____ 1. What is anything that helps you do a task called?
   a. a microscope
   b. a cylinder
   c. a measure
   d. a tool

TOOLS FOR SCIENCE

_____ 2. How is the volume of water in a jar measured?
   a. with a ruler
   b. with a graduated cylinder
   c. with a tape measure
   d. with a meterstick

Tools for Seeing

Match the correct description with the correct term. Write the letter in the space provided.

_____ 3. tool for seeing

_____ 4. tool for measuring

_____ 5. tool for analyzing

   a. meterstick
   b. computer
   c. microscope

MEASUREMENT

_____ 6. How was an inch measured in England many years ago?
   a. using grains of barley
   b. using sticks and twigs
   c. using flower petals
   d. using corn stalks

_____ 7. Which of the following is another name for the metric system?
   a. the National System of Measurement
   b. the International System of Units
   c. the Universal Unit System
   d. the International System of Measures

The International System of Units

_____ **8.** What are all units of the SI based on?
   a. the number .01
   b. the number 5
   c. the number 10
   d. the number 100

Length

Match the correct definition with the correct term. Write the letter in the space provided.

_____ **9.** the basic SI unit of length

_____ **10.** the measure of the size of a surface

_____ **11.** a measure of the amount of matter in an object

_____ **12.** the amount of space that something occupies

_____ **13.** a measure of how hot or cold something is

a. volume

b. meter

c. mass

d. area

e. temperature

WRITING NUMBERS IN SCIENTIFIC NOTATION

_____ **14.** Which of the following is used to write very large numbers and very small numbers?
   a. shorthand
   b. Morse code
   c. scientific notation
   d. the alphabet

Skills Worksheet

Directed Reading A

Section: Models in Science (pp. 50–55)

Write the letter of the correct answer in the space provided.

_____ **1.** What is a representation of an object or process called?
  a. theory
  b. law
  c. model
  d. variable

_____ **2.** Why is studying a model helpful to scientists?
  a. Scientists can study something in greater detail.
  b. Variables can never be changed.
  c. The subjects of the study are harmed.
  d. Studying models is never helpful.

TYPES OF MODELS

_____ **3.** Which of the following could NOT be represented by a model?
  a. atom
  b. Earth
  c. SI unit
  d. mathematical equation

Physical Models

_____ **4.** What are physical models?
  a. models you cannot see
  b. models you can touch
  c. models you cannot understand
  d. models you cannot touch

_____ **5.** Why is a globe a better physical model of Earth than a map is?
  a. The globe is less accurate.
  b. The map is more accurate.
  c. The globe is more accurate.
  d. The globe is flat.

Mathematical Models

_____ **6.** What are mathematical models made up of?
  a. mathematical equations and data
  b. things you can touch
  c. objects that are real
  d. things you cannot describe

_____ **7.** Which of the following are needed to process complex mathematical
models?
  a. metersticks
  b. graduated cylinders
  c. thermometers '
  d. computers

Computer Models

_____ **8.** How much data can a supercomputer process?
  a. fewer than 100 calculations every second
  b. 300 trillion calculations every second
  c. only 40 calculations every second
  d. 30 trillion calculations every second

_____ **9.** Why are computers helpful in building models?
  a. Computers can perform calculations instantly.
  b. Computers process data very slowly.
  c. Models do not contain data.
  d. Models can perform calculations instantly.

Combinations of Models

_____ **10.** Which of the following statements about models is NOT true?
  a. Climate models use information about temperature.
  b. Models track variables that affect climate.
  c. Models make exact predictions about future climates.
  d. Computers run models that track climate.

PATTERNS IN NATURE

_____ **11.** What do events in nature often follow?
  a. mathematical formulas
  b. predictable patterns
  c. unpredictable patterns
  d. scientific models

_____ **12.** What is the basis of science?
  a. designing experiments
  b. observing patterns in nature
  c. using computers
  d. making mathematical models

THEORIES AND LAWS

Use the terms from the following list to complete the sentences below.

theory   law

13. A statement or equation that reliably predicts events under certain conditions

is called a(n) _____.

14. A system of ideas that explains observations and is supported by scientific

evidence is called a(n) _____.

Theories Are Supported by Scientific Observations

_____ **15.** Which of the following statements about the theory of an
Earth-centered universe is true?
a. The theory was never questioned.
b. Observations led to a new theory.
c. The theory has been forgotten.
d. Observations proved the theory true.

Laws Can Support Theories

_____ **16.** Who discovered the law of universal gravitation?
a. Stephen Hawking
b. Copernicus
c. Benjamin Franklin
d. Sir Isaac Newton

_____ **17.** Which of the following laws supports the theory of a sun-centered
solar system?
a. the law of models
b. the universal law
c. the law of universal gravitation
d. the law of physics

Limitations of Models

_____ **18.** Why are models limited?
a. Models are not simplified.
b. Models leave out information.
c. Models cannot be understood.
d. Models are never used in science.

_____ **19.** Which of the following statements about models is true?
    a. Models are not used in science.
    b. Models change because of new data.
    c. Models can never change.
    d. Models cannot be understood by scientists.

_____ **20.** Which of the following might challenge an existing model?
    a. new technology
    b. old technology
    c. an incorrect law
    d. an incorrect theory

Skills Worksheet

Directed Reading A

Section: Mapping Earth's Surface (pp. 56–63)

Write the letter of the correct answer in the space provided.

_____ **1.** What is a representation of the features of a physical body such as Earth called?

    a. cylinder

    b. law

    c. map

    d. theory

FINDING DIRECTIONS ON EARTH

_____ **2.** What can Earth's axis of rotation be used to establish?

    a. Earth's patterns in nature

    b. Earth's temperature

    c. Earth's mass

    d. Earth's reference points

USING A COMPASS

_____ **3.** What is a tool that uses Earth's natural magnetism to show direction called?

    a. compass

    b. meterstick

    c. balance

    d. scale

_____ **4.** What does the needle of a compass point toward?

    a. magnetic west pole

    b. magnetic north pole

    c. magnetic east pole

    d. magnetic south pole

| Directed Reading A *continued*

FINDING LOCATIONS ON THE EARTH

Latitude

Use the terms from the following list to complete the sentences below.

    equator   North Pole

    parallels   latitude

5. The distance north or south from the equator is called

   _____.

6. Lines of latitude are also called _____.

7. The imaginary circle halfway between the poles is called

   the _____.

8. The _____ is 90°N latitude.

Longitude

Use the terms from the following list to complete the sentences below.

    longitude   grid   prime meridian

9. The distance east and west from the prime meridian

   is called _____.

10. The line that is designated as 0° longitude is called the

   _____.

11. Lines of latitude and longitude cross to form a(n) _____

   system on globes and maps.

INFORMATION SHOWN ON MAPS

Match the correct description with the correct term. Write the letter in the space provided.

_____ **12.** list of symbols used in a map

_____ **13.** information about a map's subject

_____ **14.** relationship between distance on Earth's
           surface and distance on a map

a. title

b. legend

c. scale

| Directed Reading A *continued*

MODERN MAPMAKING

_____ **15.** Which of the following processes provides data used in many maps?
  a. modeling
  b. remote sensing
  c. remote control
  d. magnification

Use the terms from the following list to complete the sentences below.

   passive   active   remote sensing

16. A way to get information without touching an object is called

   _____.

17. Electromagnetic radiation is recorded by sensors in a(n)

   _____ remote-sensing system.

18. Electromagnetic radiation is produced by a(n) _____

   remote-sensing system.

Global Positioning System

Use the terms from the following list to complete the sentences below.

   global positioning system   geographic information system

19. A system of orbiting satellites that sends radio signals to Earth is called

   a(n) _____.

20. A computerized system that visually presents information about an area is

   called a(n) _____.

Skills Worksheet

Directed Reading A

Section: Maps in Earth Science (pp. 64–69)

TOPOGRAPHIC MAPS

Write the letter of the correct answer in the space provided.

_____ 1. What kind of map shows the surface features of Earth?
  a. topographic map
  b. climate map
  c. road map
  d. city map

_____ 2. What is the height of an object above sea level called?
  a. longitude
  b. latitude
  c. topography
  d. elevation

CONTOUR LINES

Use the terms from the following list to complete the sentences below.

steep	contour interval	relief
contour line	gentle	index contour

3. On a topographic map, a(n) _____ connects points of equal elevation.

4. The difference in elevation between one contour line and the next is called the _____.

5. The difference in elevation between the highest and lowest points on a map is called _____.

6. Contour lines that are close together show a(n) _____ slope.

7. Contour lines that are far apart show a(n) _____ slope.

8. A dark line used to make topographic maps easier to read is called a(n)

_____.

READING A TOPOGRAPHIC MAP

_____ **9.** What color are contour lines on a topographic map?
    a. black   **c.** brown
    b. blue   **d.** pink

_____ **10.** What color are buildings, roads, and railroads on a topographic map?
    a. brown   **c.** red
    b. pink   **d.** black

THE RULES OF CONTOUR LINES

_____ **11.** Which of the following statements about contour lines is NOT true?
    a. Contour lines never cross.
    b. Contour lines that cross a valley or stream are shaped like a "U."
    c. Contour line spacing depends on the ground's slope.
    d. Contour lines that cross a valley or stream are shaped like a "V."

_____ **12.** How are the tops of hills, mountains, and depressions shown on a topographic map?
    a. by open circles
    b. by open triangles
    c. by closed triangles
    d. by closed circles

GEOLOGIC MAPS

_____ **13.** What are maps that geologists make by physically walking over an area called?
    a. contour maps
    b. geologic maps
    c. index maps
    d. remote-sensing maps

Match the correct description with the correct term. Write the letter in the space provided.

_____ **14.** a map that records information including rock units and structural features

_____ **15.** a rock of a given rock type and age range

_____ **16.** the place on a map where two geologic units meet

_____ **17.** the location of breaks in rock

a. contact

b. fault

c. geologic map

d. geologic unit

Skills Worksheet

Vocabulary and Section Summary A

Tools and Measurement

VOCABULARY

In your own words, write a definition of the following terms in the space provided.

1. meter

_____

_____

2. area

_____

_____

3. mass

_____

_____

4. volume

_____

_____

5. temperature

_____

_____

SECTION SUMMARY

Read the following section summary.

- Scientists use tools to make observations, take measurements, and analyze data.

- Scientists must select the appropriate tools for their observations and experiments to take appropriate measurements.

- Scientists use the International System of Units (SI) so that they can share and compare their observations and results with other scientists.

- Scientists have determined standard ways to measure length, area, mass, volume, and temperature.

- Scientific notation is a way to express numbers that are very large or very small.

Skills Worksheet

Vocabulary and Section Summary A

Models in Science
VOCABULARY
In your own words, write a definition of the following terms in the space provided.

1. model

_____

_____

2. law

_____

_____

3. theory

_____

_____

SECTION SUMMARY
Read the following section summary.

• Scientists must choose the right type of model to study a topic.

• Physical models, mathematical models, and computer models are common types of scientific models.

• Events in nature usually follow patterns. Scientists develop theories and laws by observing these patterns.

• Theories and laws are models that describe how the universe works. Theories and laws can change as new information becomes available.

• All models have limitations, and all models can change based on new data or new technology.

Vocabulary and Section Summary A

Mapping Earth's Surface
VOCABULARY

In your own words, write a definition of the following terms in the space provided.

1. map

_____

_____

2. equator

_____

_____

3. latitude

_____

_____

4. longitude

_____

_____

5. prime meridian

_____

_____

6. remote sensing

_____

_____

| Vocabulary and Section Summary A *continued*

SECTION SUMMARY

Read the following section summary.

- A map is a representation of the features of a physical body such as Earth.

- A compass is a tool that uses the natural magnetism of Earth to show direction.

- Latitude and longitude can be used to find points on Earth's surface.

- Most maps contain a title, a scale, a legend, an indicator of direction, and a date.

- Modern mapmakers use data gathered by remote-sensing technology to make most maps.

- Remote sensing is a way to collect information about an object without being in physical contact with the object.

- The global positioning system (GPS) calculates the latitude, longitude, and elevation of locations on Earth's surface.

- Geographic information systems (GISs) are computerized systems that allow mapmakers to store and use many types of data.

Skills Worksheet

Vocabulary and Section Summary A

Maps in Earth Science
VOCABULARY

In your own words, write a definition of the following terms in the space provided.

1. topographic map

_____

_____

2. elevation

_____

_____

3. relief

_____

_____

4. geologic map

_____

_____

SECTION SUMMARY

Read the following section summary.

- Contour lines connect points of equal elevation. They are used to show the shape of landforms.

- The contour interval is determined by the size and relief of an area.

- Geologic maps are designed to show the distribution of geologic features in a given area.

- Geologic units are the most important features shown on a geologic map.

- Geologic maps also show places where geologic units meet, where rocks are folded, and where rocks are broken.

Name _____ Class _____ Date _____

Directed Reading A

Section: The Earth System (pp. 84–89)
EARTH: AN OVERVIEW
Match the correct description with the correct term. Write the letter in the space provided.

_____ 1. the part of Earth that is water

_____ 2. the part of Earth where life exists

_____ 3. the mixture of gases that surround Earth

_____ 4. the mostly solid, rocky part of Earth

a. geosphere

b. hydrosphere

c. biosphere

d. atmosphere

GEOSPHERE
The Compositional Layers

Match the correct description with the correct term. Write the letter in the space provided.

_____ 5. the central part of Earth; made of iron and nickel

_____ 6. the thin and solid outermost layer of Earth

_____ 7. the layer of rock between Earth's core and crust

a. core

b. mantle

c. crust

The Physical Layers

Use the terms from the following list to complete the sentences below.

lithosphere   asthenosphere   mesosphere

8. Earth's cold, brittle, outermost layer is called the _____.

9. Tectonic plates move slowly on a solid, plastic layer called the

_____.

10. The lower, solid layer of the mantle is called the _____.

Use the terms from the following list to complete the sentences below.

outer core   inner core

11. Liquid iron and nickel make up Earth's _____.

12. Earth's _____ is made of solid iron and nickel.

Directed Reading A *continued*

THE ATMOSPHERE

Write the letter of the correct answer in the space provided

_____ **13.** Which of the following is the atmosphere made of?
  a. salt water from the ocean
  b. invisible gases that surround Earth
  c. iron and nickel from Earth's core
  d. rocks and minerals in Earth's crust

_____ **14.** Where are most of Earth's atmospheric gases found?
  a. within 8 to 12 km of Earth's surface
  b. more than 50 km from Earth's surface
  c. about 100 km from Earth's surface
  d. less than 4 km above Earth's surface

Layers in the Atmosphere

_____ **15.** Which layer of the atmosphere do we live in?
  a. mesosphere
  b. stratosphere
  c. thermosphere
  d. troposphere

Match the correct description with the correct term. Write the letter in the space provided.

_____ **16.** the layer directly above the troposphere

_____ **17.** the coldest atmospheric layer

_____ **18.** the uppermost layer of the atmosphere

  a. mesosphere
  b. thermosphere
  c. stratosphere

Energy Flow in the Atmosphere

Write the letter of the correct answer in the space provided,

_____ **19.** Where does the main energy source that reaches Earth's surface come from?
  a. moon
  b. lightning
  c. sun
  d. volcano

_____ **20.** Which of the following causes air in the atmosphere to move?
  a. uneven heating of Earth's surface
  b. even heating of Earth's surface
  c. uneven heating of Earth's oceans
  d. even heating of Earth's oceans

| Directed Reading A *continued*

_____ **21.** Why does cold air sink and move warm air out of the way?
   a. because cold air moves faster than warm air
   b. because warm air moves faster than cold air
   c. because cold air is less dense than warm air
   d. because cold air is denser than warm air

_____ **22.** Which of the following distributes energy throughout the atmosphere?
   a. the movement of air
   b. the formation of clouds
   c. the formation of ocean waves
   d. the movement of tectonic plates

_____ **23.** What is it called when the movement of matter results in the transfer of energy?
   a. erosion
   b. convection
   c. compression
   d. deposition

THE HYDROSPHERE

_____ **24.** What is the hydrosphere made of?
   a. only Earth's fresh water
   b. only Earth's salt water
   c. only water in glaciers
   d. all of the water in, on, and around Earth

The Global Ocean

_____ **25.** How much surface area does the global ocean cover?
   a. about 400 million square kilometers
   b. about 125 million square kilometers
   c. about 335 million square kilometers
   d. about 5 million square kilometers

_____ **26.** The global ocean holds how much of Earth's water?
   a. more than 97 percent
   b. less than 97 percent
   c. exactly 75 percent
   d. less than 12 percent

Energy Flow in the Ocean

_____ 27. Why does the temperature of ocean water vary?
   a. The sun's energy heats the oceans unevenly.
   b. The sun's energy heats the oceans evenly.
   c. Ocean water gets warmer with depth.
   d. Ocean water contains too much salt.

_____ 28. Which of the following cause differences in the density of ocean water?
   a. atmospheric pressure and salt
   b. temperature differences and salt
   c. sunlight and atmospheric pressure
   d. temperature differences and sunlight

_____ 29. Any movement of matter that results from differences in density is called what?
   a. atmospheric current
   b. geospheric current
   c. convection current
   d. hydrospheric current

_____ 30. How is energy distributed in the ocean?
   a. by solar energy
   b. by atmospheric gases
   c. by clouds and snow
   d. by convection currents

THE BIOSPHERE

_____ 31. Which of the following is NOT found in the biosphere?
   a. Earth's surface
   b. the lower part of the atmosphere
   c. the upper part of the atmosphere
   d. most of the hydrosphere

Factors Necessary for Life

_____ 32. Which of the following factors is NOT needed by most plants and animals to live?
   a. moderate climate
   b. cold climate
   c. a suitable habitat
   d. liquid water

_____ **33.** Where do plants and algae get energy from?
        a. ocean water
        b. Earth's core
        c. other organisms
        d. sun

Energy and Matter Flow in the Biosphere

Use the terms from the following list to complete the sentences below.

    decomposers   photosynthesis
    carbon dioxide   biosphere

34. Energy enters the _____ as sunlight.

35. Plants change energy from the sun into chemical energy through

    _____.

36. Energy is transferred when dead organisms are consumed by

    _____.

37. Materials such as _____ are used by plants to make food.

Skills Worksheet

Directed Reading A

Section: Heat and Energy (pp. 90–97)
WHAT IS TEMPERATURE?

Write the letter of the correct answer in the space provided.

_____ 1. What is temperature a measure of?
   a. the total kinetic energy of an object's particles
   b. the volume of an object's particles
   c. the mass of an object's particles
   d. the average kinetic energy of an object's particles

Temperature and Kinetic Energy

_____ 2. Which of the following is all matter made of?
   a. hot particles
   b. cold particles
   c. constantly moving particles
   d. nonmoving particles

_____ 3. Which of the following do particles have when they are in motion?
   a. solar energy
   b. kinetic energy
   c. chemical energy
   d. nuclear energy

_____ 4. Which of the following statements about kinetic energy is NOT true?
   a. A substance's temperature depends on the kinetic energy of the substance's particles.
   b. The more kinetic energy an object's particles have, the lower the object's temperature is.
   c. The more kinetic energy an object's particles have, the higher the object's temperature is.
   d. The faster that particles move, the more kinetic energy they have.

Average Kinetic Energy of Particles

_____ 5. Why is the *average* kinetic energy of all of an object's particles measured?
   a. because the particles are the same size
   b. because the particles do not move
   c. because the particles move at the same speed
   d. because the particles move at different speeds

_____ **6.** Why would tea in a teapot and tea in a teacup be the same
temperature, even though the teapot holds more?
  a. The tea in the teapot has a higher average kinetic energy.
  b. The tea in the teacup has a higher average kinetic energy.
  c. The tea in both containers has the same number of particles.
  d. The tea in both containers has the same average kinetic energy.

THERMAL EXPANSION

_____ **7.** When does a substance's particles have more kinetic energy?
  a. when the substance's temperature increases
  b. when the substance's temperature decreases
  c. when the substance's volume increases
  d. when the substance's particles stop moving

_____ **8.** Which of the following happens when a substance's temperature
increases?
  a. The substance's particles stop moving.
  b. The substance's particles move faster and move apart.
  c. The substance's particles move slower and move together.
  d. The substance's particles move slower and move apart.

_____ **9.** What is the increase in volume that results from an increase in
temperature called?
  a. thermal reduction
  b. thermal expansion
  c. energy reduction
  d. energy expansion

_____ **10.** Which of the following is NOT an example of thermal expansion?
  a. A mercury-filled thermometer measures temperature.
  b. A hot-air balloon rises when it is filled with hot air.
  c. A hot-air balloon sinks when it is filled with hot air.
  d. An alcohol-filled thermometer measures temperature.

WHAT IS HEAT?

_____ **11.** What is heat?
  a. energy that is absorbed by a single object
  b. energy that is transferred between objects that have different
temperatures
  c. energy that is transferred between objects that have the same
temperature
  d. energy that is absorbed when an object's temperature decreases

Transferring Heat

_____ **12.** Where does heat move when it transfers between objects?

    a. from a lower-temperature object to a higher-temperature object

    b. from a higher-temperature object to a lower-temperature object

    c. from a nonmoving object to a moving object

    d. from a slow-moving object to a fast-moving object

Use the terms from the following list to complete the sentences below.

    joules   thermal energy   temperature

13. The kinetic energy of a substance's atoms is called _____.

14. Thermal energy is expressed in _____.

15. Thermal energy depends on _____ and the amount of particles in a substance.

Reaching the Same Temperature

Write the letter of the correct answer in the space provided.

_____ **16.** What happens when two objects that have different temperatures touch?

    a. Energy passes from the cooler object to the warmer object.

    b. Both objects gain thermal energy.

    c. Energy passes from the warmer object to the cooler object.

    d. Both objects lose thermal energy.

_____ **17.** What happens when two objects that have the same temperature touch?

    a. There is no net change in the thermal energy of either object.

    b. One object loses thermal energy and the other object gains thermal energy.

    c. Both objects lose a large amount of thermal energy.

    d. Both objects gain a small amount of thermal energy.

HOW IS HEAT TRANSFERRED?

_____ **18.** How does radiation differ from conduction or convection?

    a. Radiation cannot transfer energy through empty space.

    b. Radiation can transfer energy through empty space.

    c. Radiation cannot transfer heat as electromagnetic waves.

    d. Radiation must transfer energy between two objects that are in direct contact.

| **Directed Reading A** *continued*

Match the correct definition with the correct term. Write the letter in the space provided.

_____ **19.** the transfer of energy as heat through a material

_____ **20.** the transfer of energy due to the movement of matter

_____ **21.** the transfer of heat or other energy as electromagnetic waves through matter or empty space

a. convection

b. radiation

c. conduction

STATES OF MATTER

Write the letter of the correct answer in the space provided.

_____ **22.** Which of the following is NOT a state of matter?
  a. liquid
  b. color
  c. solid
  d. gas

State and Chemical Properties

_____ **23.** Which of the following does NOT affect a substance's state?
  a. speed of particles
  b. attraction between particles
  c. color of particles
  d. pressure around particles

Changes of State

Match the correct description with the correct term. Write the letter in the space provided.

_____ **24.** changing from gas to liquid

_____ **25.** changing from liquid to solid

_____ **26.** changing from liquid to gas

a. evaporating

b. condensing

c. freezing

Name _____ Class _____ Date _____

Directed Reading A

Section: The Cycling of Energy (pp. 98–103)
THE FLOW OF ENERGY
Use the terms from the following list to complete the sentences below.

heat flow   sun
energy   waves

1. The transfer of energy from a warmer object to a cooler object is called

   _____.

2. Energy is transferred by different kinds of _____.

3. Objects carry _____ as they move.

4. A major source of energy for the Earth system is the

   _____.

RADIATION
Write the letter of the correct answer in the space provided.

_____ 5. How much energy does the sun transmit to Earth by radiation?
   a. 99%
   b. 25%
   c. 50%
   d. 5%

The Electromagnetic Spectrum
Match the correct description with the correct term. Write the letter in the space provided.

_____ 6. the type of energy that Earth receives from the sun

_____ 7. a wide range of wavelengths that includes visible light, radio waves, and gamma rays

_____ 8. a layer of Earth that absorbs energy from the sun

_____ 9. the type of energy that is transferred through Earth's systems by convection and conduction

a. atmosphere
b. thermal energy
c. electromagnetic radiation
d. electromagnetic spectrum

| Directed Reading A *continued*

CONVECTION

Write the letter of the correct answer in the space provided.

_____ **10.** How does most energy move through Earth's systems?
   a. by evaporation
   b. by conduction
   c. by radiation
   d. by convection

Use the terms from the following list to complete the sentences below.

   convection current   convection   salinity

11. The uneven heating of matter drives _____.

12. The movement of matter that results from differences in density is called

   a(n) _____.

13. Ocean water has different densities because of differences in

   _____, which is the amount of salt in water.

Convection in the Atmosphere

Use the terms from the following list to complete the sentences below.

   convection currents   mantle
   geosphere   atmosphere

14. Convection currents form in the _____ when cold air
   sinks and forces warm air away from Earth's surface.

15. Energy produced deep inside Earth heats rock in the

   _____.

16. Convection currents in the _____ carry heat from Earth's
   interior toward the surface.

17. The movement of tectonic plates is caused by _____ in
   the mantle.

Directed Reading A *continued*

CONDUCTION
Interaction of Particles
Write the letter of the correct answer in the space provided.

_____ **18.** Why does a warmer substance have more kinetic energy than a cooler substance?

    a. because particles in the warmer substance do not move

    b. because particles in the cooler substance move faster

    c. because particles in the warmer substance move faster

    d. because particles in the warmer substance move slower

_____ **19.** Which of the following happens when particles in a warm substance transfer energy to particles in a cooler substance?

    a. The cooler substance boils.

    b. The cooler substance becomes warmer.

    c. The cooler substance becomes colder.

    d. The cooler substance freezes.

Conduction Between Systems

_____ **20.** How can energy be transferred between the geosphere and atmosphere?

    a. by conduction

    b. by radiation

    c. by convection

    d. by convection currents

_____ **21.** When does the ground transfer energy to the atmosphere?

    a. when Earth's surface is warmer than the geosphere

    b. when Earth's surface is colder than the geosphere

    c. when Earth's surface is colder than the atmosphere

    d. when Earth's surface is warmer than the atmosphere

Use the terms from the following list to complete the sentences below.

    energy   conduction   air

22. Energy can be transferred between the geosphere and atmosphere

by _____.

23. When _____ touches Earth's warm surface, energy is passed to the atmosphere by conduction.

24. If the atmosphere is warmer than Earth's surface, _____ flows from the atmosphere to Earth.

| **Directed Reading A** *continued*

EARTH'S ENERGY BUDGET

Use the terms from the following list to complete the sentences below.

    open systems   spheres   energy budget

25. Energy on Earth moves through and between four _____.

26. The four spheres of Earth are _____ that constantly exchange energy with one another.

27. The movement of energy between Earth's spheres is part of Earth's

    _____.

Name _____ Class _____ Date _____

Directed Reading A

Section: The Cycling of Matter (pp. 104–111)
THE CHANGING EARTH
Write the letter of the correct answer in the space provided.

_____ 1. Which of the following statements about Earth's cycles is NOT true?
   a. Some changes on Earth happen quickly.
   b. Some changes on Earth take millions of years.
   c. Earth's surface constantly changes.
   d. Earth's surface never changes.

THE ROCK CYCLE
Match the correct description with the correct term. Write the letter in the space provided.

_____ 2. the process by which rock is broken down by wind, water, and temperature changes

_____ 3. the processes by which rocks change from one form to another

_____ 4. the process by which wind, water, ice, and gravity transport rock

a. rock cycle
b. weathering
c. erosion

Pathways in the Rock Cycle

_____ 5. Which of the following determines the pathway a rock will follow in the rock cycle?
   a. the weight of the rock
   b. the size of the rock
   c. the color of the rock
   d. the forces that act on the rock

Forces That Change Rock

_____ 6. Which of the following determines the forces that will act on a rock?
   a. the rock's color
   b. the rock's location
   c. the rock's age
   d. the rock's size

_____ **7.** Rock exposed to weathering and erosion at Earth's surface might
become which of the following?
a. sedimentary rock
b. metamorphic rock
c. igneous rock
d. inorganic rock

_____ **8.** Rock deep inside Earth is exposed to which of the following forces?
a. wind and ice
b. weathering and erosion
c. high heat and pressure
d. gravity and water

CLASSES OF ROCKS

Use the terms from the following list to complete the sentences below.

metamorphic rocks   sedimentary rocks   igneous rocks

9. Small pieces of rock that become cemented together form

_____.

10. Hot, liquid rock called magma cools and forms _____.

11. Rock that undergoes chemical changes or changes in temperature or

pressure become _____.

Sedimentary Rocks

Use the terms from the following list to complete the sentences below.

chemical   clastic
organic   sediment

12. Rock or mineral fragments that are not yet cemented together are

called _____.

13. Sediments that are buried, put under pressure, or cemented by minerals are

called _____ sedimentary rocks.

14. Minerals that crystallize from ocean water, are buried, and cemented form

_____ sedimentary rock.

15. Skeletons of dead marine animals that are buried and cemented form

_____ sedimentary rock.

Igneous Rocks

Write the letter of the correct answer in the space provided.

_____ **16.** Which of the following characteristics is NOT used to classify an igneous rock?
   a. the rock's chemical composition
   b. the rock's texture
   c. the size of the rock's crystals
   d. the rock's weight

_____ **17.** Which of the following characteristics determines the chemical composition of an igneous rock?
   a. the weight of the rock
   b. the amount of sediment in the rock
   c. the rate at which the rock cooled
   d. the type of rock that initially melts

Metamorphic Rocks

_____ **18.** Which of the following processes does NOT result in the formation of metamorphic rock?
   a. intense heat
   b. pressure
   c. weathering
   d. chemical processes

_____ **19.** Where do most metamorphic rocks form?
   a. on Earth's surface
   b. 2 km above Earth's surface
   c. on the ocean floor
   d. deep within Earth's crust

Match the correct description with the correct term. Write the letter in the space provided.

_____ **20.** has minerals that are arranged in planes or bands; gneiss

a. nonfoliated metamorphic rock

_____ **21.** has minerals that are not arranged in planes or bands; marble

b. foliated metamorphic rock

| Directed Reading A *continued*

THE WATER CYCLE

Write the letter of the correct answer in the space provided.

_____ **22.** What is the water cycle?
   a. the continuous movement of water between the atmosphere, the land, and the oceans
   b. the occasional movement of water between rivers and oceans
   c. the change in state of water from ice to liquid
   d. the occasional movement of water between the soil and plants

_____ **23.** Which of the following is the major energy source that powers the water cycle?
   a. gravity
   b. sun
   c. wind
   d. electricity

Steps of the Water Cycle

Match the correct description with the correct term. Write the letter in the space provided.

_____ **24.** the process by which liquid water changes into gaseous water vapor

_____ **25.** the process by which water vapor is released from plants into the air

_____ **26.** the process by which a gas, such as water vapor, turns into liquid water droplets

_____ **27.** the process by which water droplets fall to Earth as rain

a. condensation

b. transpiration

c. precipitation

d. evaporation

Pathways of the Water Cycle

Write the letter of the correct answer in the space provided.

_____ **28.** Where does most precipitation fall?
   a. into rivers and lakes
   b. onto dry land
   c. into the ocean
   d. over mountains

_____ **29.** What is runoff?
   a. water that runs into streams
   b. water that moves over the land surface
   c. water that freezes and becomes a glacier
   d. water that falls into the ocean

Directed Reading A *continued*

_____ **30.** What is water that moves downward through spaces in rock or soil called?
   a. groundwater
   b. precipitation
   c. overflow
   d. water vapor

THE CARBON CYCLE

_____ **31.** Where does carbon move in the Earth system?
   a. only between algae and plants
   b. only between animals and humans
   c. between ocean currents and convection currents
   d. between the nonliving environment and living things

Short-Term Processes

_____ **32.** Which of the following is NOT a way that carbon dioxide (CO_2) moves rapidly through the carbon cycle?
   a. Plants use CO_2 to build plant material.
   b. CO_2 is produced by proteins and fats.
   c. CO_2 is returned to the air after animals break down food.
   d. CO_2 is reused by plants.

_____ **33.** The process by which organisms break down dead organisms is called what?
   a. transpiration
   b. evaporation
   c. decomposition
   d. combustion

Long-Term Processes

_____ **34.** Which of the following form when dead organisms chemically change as they are compacted for millions of years?
   a. basalt and kerosene
   b. schist and gasoline
   c. granite and thermal energy
   d. limestone and fossil fuels

_____ **35.** How does carbon dioxide return to the atmosphere after fossil fuels are burned?
   a. by respiration
   b. by decomposition
   c. by combustion
   d. by convection

| Directed Reading A *continued*

THE NITROGEN CYCLE

_____ **36.** Which of the following is NOT a step in the nitrogen cycle?
    a. Bacteria change N_2 into nitrogen that plants can use.
    b. Decomposers release nitrogen from dead organisms into the soil.
    c. Plants absorb nitrogen from the air.
    d. Organisms get nitrogen by eating plants.

THE PHOSPHORUS CYCLE

_____ **37.** How do animals obtain phosphorus?
    a. by eating plants
    b. by breathing
    c. by sitting in the sun
    d. by swimming in the ocean

OTHER CYCLES IN NATURE

_____ **38.** What happens when a living thing dies?
    a. Every substance in its body is recycled.
    b. Only water in its body is recycled.
    c. Only one substance in its body is recycled.
    d. No substance in its body is recycled.

_____ **39.** Which of the following statements about cycles in nature is NOT true?
    a. Each cycle is connected to other cycles.
    b. Living things depend on cycles in nature for survival.
    c. Living things do not pass through cycles in nature.
    d. Nutrients pass between soil, plants, and animals.

Vocabulary and Section Summary A

The Earth System
VOCABULARY
In your own words, write a definition of the following terms in the space provided.

1. crust

_____

_____

2. mantle

_____

_____

3. core

_____

_____

4. convection

_____

_____

5. convection current

_____

_____

SECTION SUMMARY
Read the following section summary.

- The four divisions of Earth are the hydrosphere, atmosphere, geosphere, and biosphere.
- The geosphere is divided into layers based on composition and physical properties.
- Convection moves energy through the atmosphere and through the hydrosphere.
- Energy in the biosphere is transferred from the sun to plants and then from one organism to another.

Skills Worksheet

Vocabulary and Section Summary A

Heat and Energy
VOCABULARY
In your own words, write a definition of the following terms in the space provided.

1. temperature

_____

2. heat

_____

3. thermal energy

_____

4. conduction

_____

5. convection

_____

6. radiation

_____

SECTION SUMMARY
Read the following section summary.

• Heat moves from warmer objects to cooler objects until all of the objects are at the same temperature.

• Conduction is the transfer of energy as heat through a solid material.

• Convection is the transfer of energy due to the movement of matter.

• Radiation is the transfer of energy as electromagnetic waves. Radiation differs from conduction and convection because radiation can transfer energy through empty space.

• A substance's state of matter depends on the speed of the particles in the substance. Changes of state result from the transfer of energy.

Vocabulary and Section Summary A

The Cycling of Energy

VOCABULARY

In your own words, write a definition of the following terms in the space provided.

1. heat flow

_____

_____

_____

2. electromagnetic spectrum

_____

_____

_____

SECTION SUMMARY

Read the following section summary.

• Energy can be transferred from one place to another by heat flow, by waves, or by objects that are moving.

• Heat flow is the transfer of energy from a warmer object to a cooler object.

• Energy from the sun reaches Earth by radiation.

• Energy is transferred through the oceans, the atmosphere, and the geosphere by convection.

• Energy is transferred between the geosphere and the atmosphere by conduction.

Skills Worksheet

Vocabulary and Section Summary A

The Cycling of Matter
VOCABULARY

1. rock cycle

_____

_____

2. water cycle

_____

_____

3. carbon cycle

_____

_____

4. nitrogen cycle

_____

_____

SECTION SUMMARY

Read the following section summary.

• The processes that cycle matter in the Earth system can be relatively rapid or may take millions of years.

• The rock cycle is the series of processes in which rock changes from one form to another by geologic processes.

• The three major classes of rocks are sedimentary, igneous, and metamorphic.

• Water moves continuously from the ocean, to the atmosphere, to land, and back to the ocean through the water cycle.

• In the carbon cycle, carbon is cycled in both rapid processes and slow processes.

• Types of matter that are cycled through the Earth system include carbon, phosphorus, and nitrogen.

Directed Reading A

Section: Natural Resources (pp. 128–131)
EARTH'S RESOURCES
Write the letter of the correct answer in the space provided.

_____ **1.** What is any natural material, such as air, that is used by humans called?
   a. natural resource
   b. alternative resource
   c. human resource
   d. conservation resource

_____ **2.** Where does the energy from gasoline and wind come from?
   a. the oceans
   b. the sun
   c. the moon
   d. Earth's crust

Renewable Resources

_____ **3.** What is a resource that can be replaced at the same rate at which it is used called?
   a. recyclable resource
   b. nonrenewable resource
   c. renewable resource
   d. Earth resource

_____ **4.** Which of the following is a renewable resource?
   a. tree
   b. petroleum
   c. natural gas
   d. coal

Nonrenewable Resources

_____ **5.** What is a resource that forms more slowly than the rate at which it is used called?
   a. renewable resource
   b. Earth resource
   c. recyclable resource
   d. nonrenewable resource

_____ **6.** Which of the following is a nonrenewable resource?
   a. tree
   b. coal
   c. water
   d. air

CONSERVING NATURAL RESOURCES

_____ **7.** Which of the following is a way to conserve natural resources?
   a. Do not pollute lakes and rivers.
   b. Grow your own crops.
   c. Let water run when not in use.
   d. Save old magazines.

Energy Conservation: Reducing the Use

_____ **8.** Which of the following is a good way to conserve energy?
   a. Always use a car.
   b. Don't fill the washing machine.
   c. Leave the lights on.
   d. Take the bus.

Reusing and Recycling Resources

_____ **9.** What is reusing waste or scrap materials called?
   a. reducing
   b. refreshing
   c. recycling
   d. resourcing

_____ **10.** Which of the following is an advantage of recycling?
   a. Recycling increases the amount of trash.
   b. Recycling uses more energy than making new products.
   c. Recycling conserves energy.
   d. Recycling increases use of natural resources.

_____ **11.** Which of the following CANNOT be recycled?
   a. newspaper
   b. food
   c. aluminum can
   d. cardboard box

Skills Worksheet

Directed Reading A

Section: Rock and Mineral Resources (pp. 132–137)
MINERALS AND ROCKS
Write the letter of the correct answer in the space provided.

_____ 1. What is an inorganic solid that has a crystalline structure called?
    a. polymer
    b. mineral
    c. coal
    d. volcanic glass

_____ 2. Where do minerals form?
    a. in Earth's crust
    b. in Earth's core
    c. in chemicals
    d. in volcanic glass

_____ 3. Which of the following makes up most of the solid part of Earth's surface?
    a. silver
    b. gold
    c. rock
    d. mineral

Match the correct description with the correct term. Write the letter in the space provided.

_____ 4. Salt water dries up.

_____ 5. Pressure and temperature changes cause minerals to form.

_____ 6. Surface water carries dissolved materials into lakes.

    a. deposition
    b. evaporation
    c. metamorphism

Characteristics of Minerals
Write the letter of the correct answer in the space provided.

_____ 7. How many characteristics are used to identify a mineral?
    a. five
    b. four
    c. six
    d. two

| Directed Reading A *continued*

_____ **8.** Why is coal not a mineral?
    a. because it is a synthetic material
    b. because it is black
    c. because it has crystals
    d. because it forms from the remains of plants

MINING

_____ **9.** Why are rocks and minerals mined from the ground?
    a. so we can build quarries
    b. because they cause pollution
    c. because there are too many
    d. so they can be made into objects

_____ **10.** What are deposits that are mined for profit called?
    a. glass
    b. surface minerals
    c. quarries
    d. ore

Surface Mining

_____ **11.** What is the purpose of surface mining?
    a. to dig tunnels
    b. to protect the habitats of plants
    c. to build shafts
    d. to remove ore from Earth's surface

_____ **12.** Which of the following is NOT a type of surface mine?
    a. shaft   **c.** open pit
    b. quarry   **d.** strip mine

Subsurface Mining

_____ **13.** What must be dug when ore is too deep within Earth?
    a. deposits
    b. tunnels
    c. open pits
    d. coal

Responsible Mining

_____ **14.** Which of the following is a problem that mining may create?
    a. reclamation
    b. mineral deposits
    c. energy conservation
    d. water pollution

_____ **15.** How do we return land used for mining to its original state?
   a. by pollution
   b. by recycling
   c. by reclamation
   d. by strip mining

MAKING COMMON OBJECTS

_____ **16.** Which of the following minerals can be used as it is?
   a. diamond
   b. bauxite
   c. gypsum
   d. calcite

Metals

_____ **17.** Which of the following is NOT true of metals?
   a. Metals let light pass through.
   b. Metals are good conductors of heat.
   c. Metals have shiny surfaces.
   d. Metals are good conductors of electricity.

_____ **18.** Which of the following is made with metals?
   a. plaster of Paris
   b. fertilizer
   c. aircraft
   d. cement

Nonmetals

_____ **19.** Which of the following is NOT true of nonmetals?
   a. Nonmetals may be translucent to light.
   b. Nonmetals are good conductors of electricity.
   c. Nonmetals are good insulators of heat.
   d. Nonmetals have shiny or dull surfaces.

_____ **20.** Which of the following is a nonmetallic mineral that helps make cement?
   a. diamond
   b. coal
   c. gypsum
   d. calcite

_____ **21.** Which of the following is used to make fertilizer?
   a. diamond
   b. gypsum
   c. calcite
   d. silica

Skills Worksheet

Directed Reading A

Section: Using Material Resources (pp. 138–143)

Write the letter of the correct answer in the space provided.

_____ 1. Where do all of the objects you need to live come from?
   a. Earth
   b. atmosphere
   c. organisms
   d. sun

RESOURCES FROM EARTH

_____ 2. What kind of natural resources are used to generate energy?
   a. artificial resources
   b. energy resources
   c. material resources
   d. food resources

_____ 3. What kind of resources do humans use to make objects?
   a. material resources
   b. artificial resources
   c. atmospheric resources
   d. energy resources

Resources from the Atmosphere

_____ 4. Which of the following might be the most important resource in the atmosphere?
   a. iron
   b. argon
   c. nitrogen
   d. oxygen

_____ 5. How is argon used?
   a. as a fertilizer
   b. inside light bulbs
   c. to burn rocket fuel
   d. to make sand

Rock and Mineral Resources

_____ 6. What is iron used to make?
   a. polymers
   b. argon
   c. steel
   d. copper wiring

| Directed Reading A *continued*

_____ **7.** Where is salt harvested from?
    a. the atmosphere
    b. organisms
    c. sea water
    d. lakes

Petroleum

Use the terms from the following list to complete the sentences below.

    plastics   petroleum

8. A liquid mixture of hydrocarbons used as a fuel is called

_____.

9. Polymers, which are made from chemicals, are also called

_____.

RESOURCES FROM LIVING THINGS
Plant Resources

Match the correct description with the correct term. Write the letter in the space provided.

_____ **10.** a source of energy for humans   **a.** fruits and nuts

_____ **11.** a source of lumber and paper   **b.** cotton

                                   c. trees

_____ **12.** a source of clothing

Animal Resources

Write the letter of the correct answer in the space provided.

_____ **13.** Animals are NOT used to make which of the following products?
    a. clothing
    b. food
    c. leather
    d. cotton

_____ **14.** Which of the following animal resources can be used to fertilize crops?
    a. wastes
    b. fibers
    c. dairy products
    d. leather

THE COSTS OF MATERIAL RESOURCES
Economic Costs

_____ **15.** What is the total cost of making a product called?
   a. pollution cost
   b. natural cost
   c. environmental cost
   d. economic cost

Environmental Costs

_____ **16.** What is the damage to the land caused by making a product called?
   a. pollution cost
   b. natural cost
   c. environmental cost
   d. economic cost

_____ **17.** Which of the following may help manufacturers lower an environmental cost?
   a. pollution control
   b. marketing
   c. transport
   d. feeding

Using Resources Wisely

_____ **18.** What is one way to lower the economic cost of using a natural resource?
   a. controlling pollution
   b. improving manufacturing
   c. having faster transportation
   d. using more resources

_____ **19.** What is one way to lower the environmental cost of using a natural resource?
   a. investing in marketing
   b. using cheaper resources
   c. restoring habitats
   d. using more resources

Vocabulary and Section Summary A

Natural Resources

VOCABULARY

In your own words, write a definition of the following terms in the space provided.

1. natural resource

_____

_____

2. renewable resource

_____

_____

3. nonrenewable resource

_____

_____

4. recycling

_____

_____

SECTION SUMMARY

Read the following section summary.

- We use natural resources such as fresh water, petroleum, and trees to make our lives easier and more comfortable.

- Renewable resources can be replaced in a relatively short time, but nonrenewable resources may take thousands or even millions of years to form.

- Natural resources can be conserved by using only what is needed, by taking care of resources, and by reusing and recycling.

Skills Worksheet

Vocabulary and Section Summary A

Rock and Mineral Resources

VOCABULARY

In your own words, write a definition of the following terms in the space provided.

1. mineral

_____

_____

_____

2. ore

_____

_____

_____

SECTION SUMMARY

Read the following section summary.

- A mineral is a naturally formed, inorganic solid that has a definite crystalline structure and a consistent chemical composition.
- Environments in which minerals form may be located at or near Earth's surface or deep below the surface.
- Two types of mining are surface mining and subsurface mining.
- Two ways to reduce the harmful effects of mining are through the reclamation of mined land and the recycling of mineral products.
- Both metals and nonmetals are used to make common objects.

Vocabulary and Section Summary A

Using Material Resources

VOCABULARY

In your own words, write a definition of the following terms in the space provided.

1. material resource

_____

_____

_____

2. petroleum

_____

_____

_____

SECTION SUMMARY

Read the following section summary.

- Resources from Earth include gases from the atmosphere and rocks, minerals, and petroleum from Earth's crust.

- Living things provide humans with materials, such as food, clothing, and shelter.

- Using natural resources involves both economic and environmental costs.

- Reducing the environmental cost of using resources sometimes involves increasing the economic cost.

Skills Worksheet

Directed Reading A

Section: Fossil Fuels (pp. 158–167)

Write the letter of the correct answer in the space provided.

_____ 1. What are natural resources that humans use to generate energy called?
- **a.** Earth resources
- **b.** energy resources
- **c.** human resources
- **d.** acid resources

_____ 2. What is a nonrenewable resource formed from the remains of plants and animals called?
- **a.** Earth resource
- **b.** energy resource
- **c.** natural energy
- **d.** fossil fuel

_____ 3. Which of the following is NOT a fossil fuel?
- **a.** natural gas
- **b.** wood
- **c.** coal
- **d.** petroleum

FOSSIL FUELS AS ENERGY RESOURCES

_____ 4. When fossil fuels are burned, how is most of the energy released?
- **a.** as heat
- **b.** as petroleum
- **c.** as light
- **d.** as coal

_____ 5. Why are fossil fuels the most commonly used energy resource?
- **a.** because fossil fuels are expensive
- **b.** because fossil fuels are a renewable resource
- **c.** because fossil fuels do not burn
- **d.** because fossil fuels are inexpensive

_____ 6. How long does it take for fossil fuels to form once they have been burned?
- **a.** less than one year
- **b.** millions of years
- **c.** only 10 years
- **d.** less than 100 years

TYPES OF FOSSIL FUELS

_____ **7.** What element are all fossil fuels made up of?
   a. oxygen
   b. potassium
   c. radon
   d. carbon

_____ **8.** What does most of the carbon in fossil fuels exist as?
   a. carbohydrates
   b. fluorocarbons
   c. hydrocarbons
   d. carbohydrons

_____ **9.** Which of the following is NOT a form a fossil fuel can exist in?
   a. solid
   b. plasma
   c. liquid
   d. gas

Liquid Fossil Fuels: Petroleum

_____ **10.** What is a liquid mix of hydrocarbon compounds called?
   a. natural gas
   b. petroleum
   c. coal
   d. hydrogen

_____ **11.** How much of the world's energy comes from petroleum products?
   a. less than 10%
   b. less than 20%
   c. only 30%
   d. more than 40%

Gaseous Fossil Fuels: Natural Gas

_____ **12.** What is a gaseous mixture of hydrocarbons called?
   a. natural gas
   b. petroleum
   c. coal
   d. hydrogen

_____ **13.** What is most natural gas used for?
   a. electricity
   b. jet fuel
   c. automobile fuel
   d. heating

| Directed Reading A *continued*

_____ **14.** Which of the following is an advantage to using natural gas?
     a. Natural gas is flammable.
     b. Natural gas causes less air pollution than petroleum.
     c. Natural gas can easily cause fires.
     d. Natural gas can explode.

_____ **15.** What is the main component of natural gas?
     a. methane
     b. propane
     c. butane
     d. petroleum

Solid Fossil Fuels: Coal

_____ **16.** A solid fossil fuel formed from partially decomposed plant material is called what?
     a. natural gas
     b. petroleum
     c. coal
     d. hydrogen

_____ **17.** What is one reason less coal is used today?
     a. Coal causes pollution.
     b. Most coal has been used up.
     c. Coal is not liquid.
     d. Coal is very flammable.

_____ **18.** How is most coal used today?
     a. to generate electricity
     b. to heat homes
     c. for cars and buses
     d. for outdoor grills

HOW DO FOSSIL FUELS FORM?
Formation of Petroleum and Natural Gas

_____ **19.** From what do petroleum and natural gas mostly form?
     a. the remains of swamp plants
     b. the remains of land animals
     c. the remains of sea organisms
     d. the remains of burned-out forests

_____ **20.** What are rocks that let petroleum and gas move through them called?
     a. permeable rocks
     b. igneous rocks
     c. magma
     d. impermeable rocks

Formation of Coal

Match the correct description with the correct term. Write the letter in the space provided.

_____ **21.** brown crumbly matter used as fuel

_____ **22.** coal that is about 70% carbon

_____ **23.** coal that is about 80% carbon

_____ **24.** coal this is about 90% carbon

a. bituminous coal

b. lignite

c. peat

d. anthracite

WHERE FOSSIL FUELS ARE FOUND

Write the letter of the correct answer in the space provided.

_____ **25.** How much petroleum does the United States get from other countries?
   a. one-fourth
   b. over one-half
   c. none
   d. all

HOW FOSSIL FUELS ARE OBTAINED

_____ **26.** How do people remove petroleum and natural gas from Earth?
   a. strip mining
   b. surface mining
   c. drilling wells
   d. coal mining

_____ **27.** Which of the following processes removes soil and rock to reveal underlying coal deposits?
   a. drilling
   b. deep-sea mining
   c. shafting
   d. surface mining

PROBLEMS WITH FOSSIL FUELS

_____ **28.** Which of the following is a negative effect of using fossil fuels?
   a. radioactive waste
   b. natural resource reserves
   c. acid precipitation
   d. less pollution

_____ **29.** What is rain, sleet, or snow with a high concentration of acids called?

    a. smog

    b. acid precipitation

    c. thermal pollution

    d. environmental acid

Use the terms from the following list to complete the sentences below.

    smog   coal   oil

30. Wildlife habitats can be destroyed by surface mining for

_____.

31. When a carrier sinks, environmental problems are caused by spilled

_____.

32. Photochemical haze that forms when sunlight acts on burning fuels is

called _____.

Directed Reading A

Section: Alternative Energy (pp. 166–173)

NUCLEAR ENERGY

Use the terms from the following list to complete the sentences below.

fission   nuclear energy

1. When the nuclei of atoms are split or combined, _____

is released.

2. The process in which nuclei of radioactive atoms split into two

or more smaller nuclei is called _____.

Advantages and Disadvantages of Fission

Write the letter of the correct answer in the space provided.

_____ **3.** What is one advantage of using nuclear power rather than fossil fuels?
    a. Nuclear power does not cause air pollution.
    b. Nuclear power causes air pollution.
    c. Nuclear power results in massive strip mines.
    d. Nuclear power causes loss of wildlife habitat.

_____ **4.** Which of the following is a drawback of nuclear power plants?
    a. They produce radioactive wastes.
    b. They release smog.
    c. They release hydrocarbons.
    d. They destroy magnetic fields.

Combining Atoms: Fusion

_____ **5.** What is the joining of two or more nuclei to form a larger nucleus called?
    a. atom splitting
    b. atom splicing
    c. fission
    d. fusion

_____ **6.** Where does fusion happen naturally?
    a. in the ocean
    b. in nuclear power plants
    c. in the sun
    d. on the moon

| Directed Reading A *continued*

Advantages and Disadvantages of Fusion

_____ **7.** What is one advantage of fusion?
   a. Fusion produces few dangerous wastes.
   b. Fusion is difficult to control.
   c. Fusion needs very low temperatures.
   d. Fusion is a nonrenewable resource.

_____ **8.** What is the main disadvantage of fusion?
   a. Fusion produces air pollution.
   b. Fusion needs very high temperatures.
   c. Fusion creates a magnetic field.
   d. Fusion produces many dangerous wastes.

WIND ENERGY

_____ **9.** What is the use of a windmill to drive an electric generator called?
   a. solar power
   b. wind power
   c. hydroelectric power
   d. nuclear power

_____ **10.** Why can't all areas use wind power for electrical energy?
   a. The wind isn't strong enough.
   b. There isn't enough sunlight.
   c. Wind power is too expensive.
   d. Wind power causes too much pollution.

CHEMICAL ENERGY FROM FUEL CELLS

Use the terms from the following list to complete the sentences below.

   water   pollution
   fuel cell   chemical energy

11. Energy released when a chemical compound reacts to produce

   a new compound is called _____.

12. An energy source that changes chemical energy into electrical

   energy is a(n) _____.

13. In a fuel cell, hydrogen and oxygen react to make _____.

14. One advantage of using fuel cells for energy is that they don't

   create _____.

SOLAR ENERGY

Use the terms from the following list to complete the sentences below.

solar cells   solar energy   sun

15. Almost all forms of energy come from the _____.

16. Radiation energy from the sun is called _____.

17. Sunlight can be turned into electrical energy by using

_____

HYDROELECTRIC ENERGY

Write the letter of the correct answer in the space provided.

_____ **18.** How is hydroelectric energy produced?
    a. from solar energy
    b. by moving water
    c. through fuel cells
    d. by fusion

Advantages of Hydroelectric Energy

_____ **19.** What is one advantage of hydroelectric energy?
    a. Hydroelectric energy is renewable.
    b. Hydroelectric energy can be used everywhere.
    c. Hydroelectric energy creates wildlife habitats.
    d. Hydroelectric energy helps migrating birds.

Disadvantages of Hydroelectric Energy

_____ **20.** What is one disadvantage of building a dam for a hydroelectric
power plant?
    a. A dam can disrupt wildlife.
    b. A dam helps fish migrate.
    c. A dam improves water quality.
    d. A dam will never collapse.

ENERGY FROM LIVING THINGS

_____ **21.** What is organic matter that is used as an energy source called?
    a. biomass
    b. atoms
    c. hydrocarbons
    d. nuclei

_____ **22.** What can plants that contain sugar or starch be made into?
    a. gasoline
    b. alcohol
    c. methane
    d. hydrocarbons

_____ **23.** What fuel is made by mixing alcohol and gasoline?
    a. coal
    b. petroleum
    c. gasohol
    d. biomass

Burning Biomass

_____ **24.** When biomass is burned, how is energy released?
    a. as heat
    b. as wind
    c. as water
    d. as alcohol

Advantages and Disadvantages of Biomass

_____ **25.** What is one advantage of using biomass for fuel?
    a. Biomass is inexpensive.
    b. Biomass does not burn.
    c. Biomass grows back slowly.
    d. Biomass is expensive.

_____ **26.** What is one disadvantage of using biomass for fuel?
    a. Using biomass too quickly may destroy habitats.
    b. Growing biomass is difficult.
    c. Biomass is inorganic.
    d. Burning biomass does not get hot.

ENERGY FROM WITHIN EARTH

_____ **27.** What is magma?
    a. solid rock
    b. melted rock
    c. steam
    d. pollution

_____ **28.** What is energy produced by heat inside Earth called?
    a. geographic energy
    b. volcanic energy
    c. nuclear energy
    d. geothermal energy

Directed Reading A *continued*

Geothermal Energy

_____ **29.** What turns the turbines in a geothermal power plant?
  a. vents and generators
  b. hot water and steam
  c. groundwater and magma
  d. solid rock and magma

Advantages and Disadvantages of Geothermal Energy

_____ **30.** What is one advantage of using geothermal energy?
  a. It causes a lot of pollution.
  b. It doesn't require a power plant.
  c. It can be used anywhere.
  d. It is renewable.

Skills Worksheet

Vocabulary and Section Summary A

Fossil Fuels
VOCABULARY

In your own words, write a definition of the following terms in the space provided.

1. energy resource

_____

_____

_____

2. fossil fuel

_____

_____

_____

3. petroleum

_____

_____

_____

4. natural gas

_____

_____

_____

5. coal

_____

_____

_____

6. acid precipitation

_____

_____

_____

SECTION SUMMARY

Read the following section summary.

- Energy resources are natural resources that humans use to produce energy.

- Fossil fuels are nonrenewable resources that form slowly over long periods of time from the remains of dead organisms. Petroleum, natural gas, and coal are fossil fuels.

- When fossil fuels are burned, they release energy. Most of that energy is heat energy.

- How humans use fossil fuels depends on the availability of the fuel, the ways in which the fuels are converted into energy, and the effects of converting the fuels into energy.

- Fossil fuels are found all over the world. The United States imports more than half of the petroleum that it uses from the Middle East, South America, Africa, Mexico, and Canada.

- Fossil fuels are obtained by drilling oil wells, mining below Earth's surface, and strip mining.

- Acid precipitation, smog, water pollution, and the destruction of wildlife habitats are some of the environmental problems created by the use of fossil fuels.

Skills Worksheet

Vocabulary and Section Summary A

Alternative Energy
VOCABULARY

In your own words, write a definition of the following terms in the space provided.

1. nuclear energy

_____

_____

_____

2. wind power

_____

_____

_____

3. chemical energy

_____

_____

_____

4. solar energy

_____

_____

_____

5. hydroelectric energy

_____

_____

_____

6. biomass

_____

_____

_____

▌Vocabulary and Section Summary A *continued*

7. gasohol

_____

_____

_____

8. geothermal energy

_____

_____

_____

SECTION SUMMARY

Read the following section summary.

- The usefulness of a resource depends on how easy converting the resource into energy is, how expensive the resource is, and how much of the resource is available.

- Fission and fusion are processes that release nuclear energy. The byproducts of fission are heat and radioactive waste.

- Wind power, solar energy, hydroelectric energy, biomass, and geothermal energy are renewable resources that emit very little pollution.

- Not all alternative energy resources can be generated in all areas. Some alternative energy resources are very expensive.

- Every energy resource has advantages and disadvantages.

Directed Reading A

Section: Earth's Structure (pp. 190–197)
THE LAYERS OF EARTH

Write the letter of the correct answer in the space provided.

_____ 1. What is Earth made of?
  a. several layers
  b. hollow space
  c. solid rock
  d. one layer

_____ 2. Scientists think about Earth's layers in terms of which of the following ways?
  a. their physical composition and their chemical properties
  b. their chemical composition and their physical properties
  c. their temperature and their composition
  d. their mass and their temperature

The Compositional Layers of Earth

Match the correct description with the correct term. Write the letter in the space provided.

_____ 3. surface layer of Earth made of silicon, oxygen, and aluminum

_____ 4. dense, thick middle layer of Earth made of silicon, oxygen, and magnesium

_____ 5. central layer of Earth made mainly of iron

a. core

b. crust

c. mantle

Continental and Oceanic Crust

Use the terms from the following list to complete the sentences below.

continental   layers
oxygen   oceanic

6. Continental crust is thicker than _____ crust.

7. Both oceanic crust and continental crust are made mainly

of _____, silicon, and aluminum.

8. Oceanic crust is heavier than _____ crust.

9. Compared with Earth's other _____, both types of crust

are rocky, thin, and fractured.

The Physical Structure of Earth

Match the correct description with the correct term. Write the letter in the space provided.

_____ **10.** number of Earth's layers based on physical properties

_____ **11.** outer layer of Earth, including the crust and upper part of the mantle

_____ **12.** layer of the mantle made of solid rock that flows slowly

_____ **13.** lower part of the mantle

_____ **14.** layer of liquid iron and nickel

_____ **15.** solid center of Earth

a. lithosphere

b. asthenosphere

c. five

d. outer core

e. mesosphere

f. inner core

MAPPING EARTH'S INTERIOR

Write the letter of the correct answer in the space provided.

_____ **16.** What causes seismic waves?
   a. winds
   b. an earthquake
   c. magnetic reversal
   d. rain

_____ **17.** What can scientists find out about Earth with a seismometer?
   a. density and thickness of Earth's layers
   b. Earth's age
   c. Earth's atmosphere
   d. Earth's temperature

_____ **18.** Which of the following affects the speed of seismic waves?
   a. Earth's temperature
   b. Earth's weather conditions
   c. the material the waves are made of
   d. the material the waves pass through

CONTINENTAL DRIFT
Restless Continents

_____ **19.** What is the hypothesis that a single large landmass broke up into smaller landmasses to form the continents, which then drifted to their present locations?
   a. continental spreading
   b. plate tectonics
   c. Wegener's puzzle
   d. continental drift

_____ **20.** Which of the following does this hypothesis explain?
   a. how the continents differ in terrain
   b. how the continents differ in climate
   c. how the continents seem to fit together
   d. how the continents are similar

Evidence for Continental Drift

_____ **21.** How do fossils help explain continental drift?
   a. Fossils show that animals crossed the ocean.
   b. Fossils of the same species are found on continents that are far from each other.
   c. Fossils show when drift happened.
   d. Fossils formed when drift happened.

THE BREAKUP OF PANGAEA

_____ **22.** What did Wegener call the single large continent?
   a. Pangaea
   b. Laurasia
   c. Gondwana
   d. Eurasia

_____ **23.** Which of the following was NOT a landform caused by the continents drifting and colliding with each other?
   a. mountain ranges
   b. volcanoes
   c. ocean trenches
   d. lakes

Directed Reading A *continued*

SEA-FLOOR SPREADING

_____ **24.** What made many scientists think the continents could not have
changed position?
a. They couldn't find fossil evidence.
b. The rocks seemed too strong.
c. Earth seemed too young.
d. The oceans seemed too widely separated.

Mid-Ocean Ridges—a Magnetic Mystery

_____ **25.** Which of the following discoveries did scientists make when studying
the ocean floor in the 1960s?
a. magma
b. mid-ocean ridges
c. magnetized rocks
d. a submerged mountain chain

Magnetic Reversals—Mystery Solved

_____ **26.** What is it called when Earth's magnetic poles change places?
a. mid-ocean reversal
b. magnetic reversal
c. polar drift
d. sea-floor reversal

Sea-Floor Spreading

_____ **27.** The record of magnetic reversals shown on the sea floor provides
evidence of what?
a. The continents are moving.
b. Mid-ocean ridges exist.
c. Ocean volcanoes happen.
d. Tectonic plates exist.

Directed Reading A

Section: The Theory of Plate Tectonics (pp. 198–203)

Write the letter of the correct answer in the space provided.

_____ 1. What is the name of the theory that Earth's lithosphere is divided into tectonic plates?
 a. plate theory
 b. tectonic theory
 c. plate tectonics
 d. convergent theory

TECTONIC PLATES

_____ 2. What are large pieces of the lithosphere that move around on top of the asthenosphere called?
 a. mantle pieces
 b. crust plates
 c. tectonic plates
 d. puzzle pieces

A Tectonic Plate Close-Up

_____ 3. How do tectonic plates fit together?
 a. like a layer cake
 b. like a jigsaw puzzle
 c. like a stack of books
 d. like a model car

_____ 4. Which of the following is the thickest part of the South American plate?
 a. the part underneath the continental crust
 b. the oceanic crust
 c. the mantle
 d. the mid-Atlantic Ocean

Like Ice Cubes in a Bowl of Punch

_____ 5. How are tectonic plates like ice cubes in a bowl of punch?
 a. Tectonic plates move around and touch each other.
 b. Tectonic plates melt and become liquid.
 c. Tectonic plates sink and disappear from the surface.
 d. Tectonic plates freeze and become harder.

| Directed Reading A *continued*

TECTONIC PLATE BOUNDARIES

_____ **6.** What is a place where tectonic plates touch called?

     a. a separation

     b. a collision

     c. a division

     d. a boundary

Convergent Boundaries

_____ **7.** What is it called when the denser of two tectonic plates sinks beneath the less dense plate after they collide?

     a. abduction

     b. subduction

     c. deduction

     d. collision

_____ **8.** What is a series of volcanic islands called?

     a. island arc

     b. island arch

     c. volcanic arc

     d. volcanic ash

Divergent Boundaries

_____ **9.** Where do most divergent boundaries happen?

     a. in Iceland

     b. in East Africa

     c. on land

     d. on the sea floor

Transform Boundaries

_____ **10.** The San Andreas fault system in California is a well-known example of what kind of boundary?

     a. convergent boundary

     b. divergent boundary

     c. transform boundary

     d. convergent and transform boundaries

_____ **11.** Where is the San Andreas fault system located?
  a. where the Pacific and the North American plates are sliding past each other
  b. where the Atlantic and the North American plates are sliding past each other
  c. where the Pacific and the South American plates are sliding past each other
  d. where the Atlantic and the South American plates are sliding past each other

Match the correct description with the correct term. Write the letter in the space provided.

_____ **12.** boundary at which two tectonic plates collide

_____ **13.** boundary at which two tectonic plates separate

_____ **14.** boundary at which two tectonic plates slide past one another horizontally

a. transform boundary

b. convergent boundary

c. divergent boundary

CAUSES OF TECTONIC PLATE MOTION

Write the letter of the correct answer in the space provided.

_____ **15.** What causes the motion of tectonic plates?
  a. differences in density
  b. changes in atmosphere
  c. changes in Earth's core
  d. changes in the oceans

_____ **16.** What happens when rock is heated?
  a. Rock breaks.
  b. Rock rises toward Earth's surface.
  c. Rock sinks.
  d. Rock becomes denser.

_____ **17.** What happens when rock cools?
  a. Rock breaks.
  b. Rock rises toward Earth's surface.
  c. Rock sinks.
  d. Rock becomes denser.

| **Directed Reading A** *continued*

Match the correct description with the correct term. Write the letter in the space provided.

_____ **18.** The edge of a tectonic plate sinks and pulls the rest of the plate with it.

_____ **19.** Gravity makes the tectonic plate slide downhill.

_____ **20.** Heating and cooling of rocks make a tectonic plate move sideways.

a. ridge push

b. convection

c. slab pull

TRACKING TECTONIC PLATE MOTION

Write the letter of the correct answer in the space provided.

_____ **21.** How is the movement of tectonic plates measured?
   a. in kilometers per year
   b. in meters per year
   c. in centimeters per year
   d. in centimeters per day

_____ **22.** What do scientists use to measure tectonic plate movement on continents?
   a. clinometers
   b. global positioning system (GPS)
   c. telescopes
   d. seismometers

_____ **23.** What do scientists use to measure the rate of movement of oceanic plates?
   a. sea-floor spreading
   b. global positioning system (GPS)
   c. seismometers
   d. magnetic reversal

Skills Worksheet

Directed Reading A

Section: Deforming Earth's Crust (pp. 204–209)
DEFORMATION
Use the terms from the following list to complete the sentences below.

deformation	bend
stress	break

1. The amount of force per unit area on a given material is called

_____.

2. The process by which the shape of a rock changes in response to stress

is called _____.

3. Rock layers may _____ when stress is placed

on them, but when enough stress is placed on rocks, the rocks

may _____.

FOLDING
Write the letter of the correct answer in the space provided.

_____ **4.** What is the bending of rock layers due to stress called?
  a. deformation   **c.** faulting
  b. folding   **d.** stress

_____ **5.** How many limbs does each fold have?
  a. one   **c.** three
  b. two   **d.** four

Match the correct description with the correct term. Write the letter in the space provided.

_____ **6.** a fold in which one limb is tilted beyond 90°

_____ **7.** a fold in which the oldest rock layers are in the center of the fold

_____ **8.** a fold in which one limb dips more steeply than the other limb does

_____ **9.** a fold in which the youngest rock layers are in the center of the fold

_____ **10.** a lying-down fold

a. recumbent fold

b. syncline

c. asymmetrical fold

d. anticline

e. overturned fold

FAULTING

Match the correct description with the correct term. Write the letter in the space provided.

_____ 11. a break in a body of rock along which one block slides relative to another

_____ 12. the blocks of crust on each side of the fault

_____ 13. the block of rock that lies below the plane of the fault

_____ 14. the block of rock that lies above the plane of the fault

a. hanging wall

b. footwall

c. fault

d. fault blocks

Normal Faults

Use the terms from the following list to complete the sentences below.

compression   reverse   strike-slip
tension   normal   shear stress

15. When the hanging wall moves down relative to the footwall, it is

a(n) _____ fault.

16. Normal faults usually form where tectonic plate motions cause

_____.

17. Along a(n) _____ fault, the hanging wall moves up relative to the footwall.

18. Reverse faults usually form where tectonic plate motions cause

_____.

19. When opposing forces cause rocks to break and move horizontally,

_____ faults form.

20. Strike-slip faults usually form where tectonic plate motions cause

_____.

Recognizing Faults

Write the letter of the correct answer in the space provided.

_____ **21.** Which of the following is NOT a likely indicator of a fault?
  a. layers of different kinds of rock that sit side-by-side
  b. scarps
  c. seismic waves
  d. slickensides

PLATE TECTONICS AND MOUNTAIN BUILDING

_____ **22.** Which of the following may cause mountain building over long periods of time?
  a. mid-ocean ridges
  b. the movement of tectonic plates
  c. recumbent folds
  d. continental drift

Folded Mountains

_____ **23.** What mountains are formed when rock layers are squeezed together and pushed upward?
  a. folded mountains
  b. fault-block mountains
  c. volcanic mountains
  d. strike-slip mountains

_____ **24.** Folded mountain ranges form at what type of boundaries?
  a. divergent boundaries
  b. convergent boundaries
  c. transform boundaries
  d. divergent and transform boundaries

Fault-Block Mountains

_____ **25.** What type of mountains are formed when tension causes large blocks of Earth's crust to drop down relative to other blocks?
  a. folded mountains
  b. fault-block mountains
  c. volcanic mountains
  d. strike-slip mountains

_____ **26.** Which of the following are a range of fault-block mountains?
  a. Appalachian Mountains
  b. Himalaya Mountains
  c. Tetons
  d. Andes

Volcanic Mountains

_____ **27.** What type of mountains are formed when molten rock erupts onto
Earth's surface?
  a. folded mountains
  b. fault-block mountains
  c. volcanic mountains
  d. strike-slip mountains

_____ **28.** Most major volcanic mountains are located at which type of
boundaries?
  a. convergent boundaries
  b. divergent boundaries
  c. transform boundaries
  d. divergent and transform boundaries

_____ **29.** Most of Earth's active volcanic mountains have formed around
the tectonically active rim of what ocean?
  a. Arctic Ocean
  b. Indian Ocean
  c. Atlantic Ocean
  d. Pacific Ocean

Name _____ Class _____ Date _____

Directed Reading A

Section: California Geology (pp. 210–217)

Write the letter of the correct answer in the space provided.

_____ 1. How long has California been at an active plate boundary?
    a. one year
    b. 100 years
    c. 100 million years
    d. 225 million years

_____ 2. What has been the most important force in California's geology?
    a. water
    b. fire
    c. plate tectonics
    d. gravity

BUILDING CALIFORNIA BY PLATE TECTONICS

_____ 3. Where was the western edge of North America about 225 million years ago compared to today?
    a. underwater
    b. on another plate
    c. farther west
    d. farther east

_____ 4. Which of the following processes began the most important part of geologic "building" in California between 160 million and 100 million years ago?
    a. convection
    b. subduction
    c. deformation
    d. separation

Ancient Plate Boundaries

_____ 5. Which of the following major tectonic plates interacted to develop the modern plate boundary between the North American and Pacific plates?
    a. North American, Farallon, and Pacific plates
    b. North American and Pacific plates
    c. North American and Farallon plates
    d. Pacific and Farallon plates

_____ **6.** Which of the following boundaries existed between the ancient North American and Farallon plates?
   a. convergent boundary
   b. divergent boundary
   c. transform boundary
   d. parallel boundary

_____ **7.** When the Farallon plate subducted beneath the North American plate, where was the subduction zone located?
   a. northern California
   b. southern California
   c. along the mid-section of California
   d. along all of California

The Transform Boundary Is Born

_____ **8.** Which of the following plates subducted to help form the transform boundary?
   a. Pacific plate
   b. Juan de Fuca plate
   c. Farallon plate
   d. North American plate

Match the correct description with the correct term. Write the letter in the space provided.

_____ **9.** plate that completely subducted at one part of the boundary about 25 million years ago

_____ **10.** what was created when the Pacific plate touched North America for the first time

_____ **11.** remains from the ancient Farallon plate

_____ **12.** plate that moved closer and closer to North America

a. Farallon plate

b. Pacific plate

c. Juan de Fuca plate

d. transform boundary

SUBDUCTION AND VOLCANISM

Write the letter of the correct answer in the space provided.

_____ **13.** Which of the following was NOT one of the results of the subduction of the Farallon plate?
   a. Rock melted.
   b. Rock collided with the North American continent.
   c. Magma was formed.
   d. The oceanic plate cooled as it subducted.

| Directed Reading A *continued*

The Sierra Nevada Batholith

_____ **14.** How long ago did the granite rocks of the Sierra Nevadas form?
     a. more than 225 million years ago
     b. more than 210 million years ago
     c. more than 100 million years ago
     d. around 25 million years ago

_____ **15.** What is a batholith?
     a. an active volcano
     b. a large mass of rock in Earth's crust
     c. a flow of lava
     d. a massive wave

The Cascadia Subduction Zone

_____ **16.** Which of the following were formed as a result of the Cascadia subduction zone?
     a. active volcanoes in the Cascade Mountains
     b. active volcanoes in the Appalachian Mountains
     c. active volcanoes in the Rocky Mountains
     d. active volcanoes in the Himalaya Mountains

_____ **17.** Where are Lassen Peak and Mount Shasta in California situated?
     a. at the northernmost point of the Cascade volcanic chain
     b. at the southernmost point of the Cascade volcanic chain
     c. at the easternmost point of the Cascade volcanic chain
     d. at the westernmost point of the Cascade volcanic chain

SUBDUCTION AND ACCRETION

_____ **18.** Accretion has helped to form which of the following?
     a. earthquakes
     b. volcanoes
     c. oceans
     d. mountain chains

Accreted Terranes

_____ **19.** Which of the following is a piece of lithosphere that becomes part of a larger landmass when tectonic plates collide?
     a. plate boundary
     b. gold
     c. accreted terrane
     d. volcano

_____ **20.** Which of the following is NOT information used by geologists to identify accreted terrane?

    a. Terrane rocks differ from the surrounding rocks.

    b. Terrane rocks are identical to the surrounding rocks.

    c. The terrane is often surrounded by faults.

    d. Terrane rocks may consist of marine sedimentary rocks.

California Gold

_____ **21.** Where is most of California's gold found?

    a. underwater

    b. in southern California

    c. in rock along the western side of the Sierra Nevadas

    d. in volcanoes

THE SAN ANDREAS FAULT SYSTEM

_____ **22.** Which of the following forms the boundary between the Pacific and North American plates?

    a. San Andreas fault

    b. accreted terrane

    c. a chain of volcanoes

    d. Cascade Mountain range

_____ **23.** In which direction does the San Andreas fault extend from the California-Mexico border to northern California?

    a. northeast

    b. northwest

    c. southeast

    d. southwest

Plate Motion on the San Andreas Fault System

_____ **24.** It is best to think of the boundary between the Pacific and North American plates as which of the following?

    a. as a square

    b. as a circle

    c. as a line

    d. as a zone

Offset on the San Andreas Fault System

_____ **25.** The offset along the San Andreas fault system is about how far?

    a. 3 km

    b. 315 km

    c. 3,000 km

    d. 30,000 km

Compression in Southern California

_____ **26.** Which of the following is happening as a result of the bend in the San Andreas fault?
 a. The Pacific and North American plates are separating.
 b. The Pacific and North American plates are sliding past each other.
 c. The Pacific and North American plates are colliding.
 d. The Pacific and North American plates are merging.

_____ **27.** What is happening to the land in southern California that is being compressed?
 a. It is being uplifted or dropped down.
 b. It is becoming flat.
 c. It is subducting.
 d. It is erupting.

PLATE TECTONICS AND THE CALIFORNIA LANDSCAPE

_____ **28.** What force helped form much of California's landscape?
 a. ocean water
 b. plate tectonics
 c. gravity
 d. fault lines

_____ **29.** What has formed central and northern California's steep, rocky coastline?
 a. uplift
 b. fault lines
 c. gravity
 d. ocean water

_____ **30.** Why does most of California's landscape have a northwesterly orientation?
 a. because northwest is perpendicular to the faults of the plate boundary
 b. because northwest is parallel to the faults of the plate boundary
 c. because northwest is transverse to the faults of the plate boundary
 d. because east-west is parallel to the faults of the plate boundary

Name _____ Class _____ Date _____

Vocabulary and Section Summary A

Earth's Structure
VOCABULARY
In your own words, write a definition of the following terms in the space provided.

1. core

_____

_____

2. mantle

_____

_____

3. crust

_____

_____

4. lithosphere

_____

_____

5. asthenosphere

_____

_____

6. continental drift

_____

_____

7. sea-floor spreading

_____

_____

| Vocabulary and Section Summary A *continued*

SECTION SUMMARY

Read the following section summary

- Earth is made up of three layers—the crust, the mantle, and the core—based on chemical composition. Of these three layers, the core is made up of the densest materials. The crust and mantle are made up of materials that are less dense than the core.

- Earth is made up of five layers—the lithosphere, the asthenosphere, the mesosphere, the outer core, and the inner core—based on physical properties.

- Knowledge about the layers of Earth comes from the study of seismic waves caused by earthquakes.

- Wegener hypothesized that continents drift apart from one another now and that they have drifted in the past.

- Magnetic reversals that occur over time are recorded in the magnetic pattern of the oceanic crust, which provides evidence of sea-floor spreading and continental drift.

- Sea-floor spreading is the process by which new sea floor forms at mid-ocean ridges.

Vocabulary and Section Summary A

The Theory of Plate Tectonics
VOCABULARY

In your own words, write a definition of the following terms in the space provided.

1. plate tectonics

_____

_____

2. tectonic plate

_____

_____

SECTION SUMMARY

Read the following section summary

- Plate tectonics is the theory that explains how pieces of Earth's lithosphere move and change shape.

- Tectonic plates are large pieces of the lithosphere that move around on top of the asthenosphere.

- Boundaries between tectonic plates are classified as convergent, divergent, or transform.

- Convection is the main driving force of plate tectonics.

- Tectonic plates move a few centimeters per year. Scientists measure this rate by using GPS or by using sea-floor spreading.

Skills Worksheet

Vocabulary and Section Summary A

Deforming Earth's Crust

VOCABULARY

In your own words, write a definition of the following terms in the space provided.

1. folding

_____

_____

2. fault

_____

_____

SECTION SUMMARY

Read the following section summary

- Deformation structures, such as faults and folds, form as a result of stress in the lithosphere. This stress is caused by tectonic plate motion.
- Folding occurs when rock layers bend because of stress.
- Faulting occurs when rock layers break because of stress and then move on either side of the break.
- Three major fault types are normal faults, reverse faults, and strike-slip faults.
- Mountain building is caused by the movement of tectonic plates. Folded mountains and volcanic mountains form at convergent boundaries. Fault-block mountains form at divergent boundaries.

Vocabulary and Section Summary A

California Geology

VOCABULARY

In your own words, write a definition of the following terms in the space provided.

1. batholith

_____

_____

2. accreted terrane

_____

_____

_____

SECTION SUMMARY

Read the following section summary

- Plate tectonics has been the most important force in the shaping of California's geology.

- When Pangaea broke apart, the western edge of North America became an active plate boundary.

- Between 225 million and 25 million years ago, subduction took place along all of California.

- During subduction, California grew larger as accreted terranes were added to the North American continent.

- A transform boundary formed about 25 million years ago, when the Pacific plate met the North American plate.

- Along the San Andreas fault, the Pacific plate is moving northwest relative to the North American plate.

- The motions of tectonic plates have caused mountains and valleys to form in California.

Name _____ Class _____ Date _____

Directed Reading A

Section: What Are Earthquakes? (pp. 232–237)
Write the letter of the correct answer in the space provided.

_____ **1.** What is a ground movement caused when rocks move and release energy called?
  a. tectonic plate   **c.** earthquake
  b. fault   **d.** mid-ocean ridge

WHERE EARTHQUAKES HAPPEN

_____ **2.** Where do earthquakes NOT take place?
  a. in continental lithosphere
  b. in the interior of the North American plate
  c. far from tectonic plate boundaries
  d. near tectonic plate boundaries

_____ **3.** What is a break in Earth's crust along which blocks of rock move called?
  a. seismic wave
  b. boundary
  c. rebound
  d. fault

FAULTS AT TECTONIC PLATE BOUNDARIES
Use the terms from the following list to complete the sentences below.

| fault zone | transform | divergent |
| convergent | subduction | |

4. At _____ boundaries, earthquakes happen along normal faults at depths of less than 20 km.

5. At _____ boundaries, earthquakes can happen along reverse faults at depths of as much as 700 km.

6. The process of one tectonic plate moving beneath another is called

_____ .

7. At _____ boundaries, earthquakes happen along strike-slip faults as blocks of rock move.

8. An area along plate boundaries where many interconnected faults are

located is called a(n) _____ .

| Directed Reading A *continued*

WHY EARTHQUAKES HAPPEN

Use the terms from the following list to complete the sentences below.

seismic waves   plastic deformation
elastic deformation   elastic rebound   stress

9. As tectonic plates move, _____ on rocks near the edges of

the plates increases.

10. When rock deforms like a piece of clay being molded,

_____ has happened.

11. When rock deforms like a rubber band being stretched,

_____ has happened.

12. The sudden return of elastically deformed rock to its undeformed shape is

called _____.

13. Earthquakes happen when energy travels through rock as

_____.

EARTHQUAKE WAVES

Use the terms from the following list to complete the sentences below.

surface waves   body waves

14. Seismic waves that travel through Earth's interior are called

_____.

15. Seismic waves that travel along Earth's surface are called

_____.

P Waves

Write the letter of the correct answer in the space provided.

_____ **16.** Which of the following waves are the fastest seismic waves?
   a. surface waves
   b. S waves
   c. P waves
   d. shear waves

_____ **17.** What are P waves often called because they are the first waves to be
   detected?
   a. surface waves
   b. primary waves
   c. secondary waves
   d. shear waves

_____ **18.** Which of the following waves travel through solids, liquids, and gases?
  a. P waves
  b. S waves
  c. surface waves
  d. secondary waves

S Waves

_____ **19.** What are the second-fastest waves called?
  a. S waves
  b. surface waves
  c. primary waves
  d. pressure waves

_____ **20.** Which of the following waves shear rock horizontally from side to side?
  a. surface waves
  b. primary waves
  c. pressure waves
  d. S waves

_____ **21.** Which of the following waves cannot travel through liquids?
  a. S waves
  b. P waves
  c. pressure waves
  d. primary waves

Surface Waves

_____ **22.** Why do surface waves tend to cause the most damage?
  a. because energy is focused above Earth's surface
  b. because energy is focused on Earth's surface
  c. because energy is focused below Earth's surface
  d. because no energy is produced

_____ **23.** How do surface waves differ from body waves?
  a. Surface waves travel faster than body waves.
  b. Surface waves travel more slowly than body waves.
  c. Surface waves are less destructive than body waves.
  d. Surface waves are no different than body waves.

_____ **24.** How many types of surface waves are there?
  a. one
  b. three
  c. two
  d. six

Directed Reading A

Section: Earthquake Measurement (pp. 238–243)
STUDYING EARTHQUAKES

Match the correct description with the correct term. Write the letter in the space provided.

_____ **1.** the point on Earth's surface directly above an earthquake's starting point

_____ **2.** a tracing of earthquake motion

_____ **3.** the location within Earth where an earthquake begins

a. seismogram
b. epicenter
c. focus

Triangulation
Write the letter of the correct answer in the space provided.

_____ **4.** What do scientists locate when they draw circles around three seismometer stations and find where the circles intersect?
a. a seismogram
b. an earthquake's epicenter
c. S waves
d. P waves

Match the correct description with the correct term. Write the letter in the space provided.

_____ **5.** the time between the arrival of P waves and S waves

_____ **6.** the process used to find an earthquake's epicenter

a. triangulation
b. lag time

| Directed Reading A *continued*

EARTHQUAKE MAGNITUDE

Use the terms from the following list to complete the sentences below.

moment magnitude scale   Richter scale
magnitude

7. The measure of an earthquake's strength is called _____.

8. Earthquake magnitude can be found by using the _____,
which measures ground motion and adjusts for distance.

9. Most scientists today consider the _____ a more accurate
measure of earthquake strength.

EARTHQUAKE INTENSITY

Write the letter of the correct answer in the space provided.

_____ **10.** What is the measurement of the effects of an earthquake at Earth's
surface called?
a. magnitude
b. focus
c. intensity
d. lag time

The Modified Mercalli Scale

_____ **11.** Which of the following scales is used to measure earthquake
intensity?
a. the Richter scale
b. the moment magnitude scale
c. the Modified Mercalli scale
d. the Earthquake Hazard scale

_____ **12.** What is the lowest intensity level on the Modified Mercalli scale?
a. I
b. X
c. XII
d. XX

_____ **13.** How much damage is caused by an earthquake with an intensity level
of XII?
a. no damage
b. severe damage
c. barely noticeable damage
d. total destruction

Mapping Earthquake Intensity

_____ 14. What are earthquake intensity maps used for?
a. to explain what causes earthquakes
b. to predict future earthquake damage
c. to find an earthquake's epicenter
d. to find fault zones

THE EFFECTS OF EARTHQUAKES

Distance from the Epicenter

Use the terms from the following list to complete the sentences below.

liquefaction   epicenter   flexible

15. Factors that determine the effects of earthquakes include distance from the

_____, local geology, and type of construction.

16. A process called _____ can intensify ground shaking or cause the ground to settle.

17. Structures made of _____ building materials are more likely to survive strong ground shaking.

Directed Reading A

Section: Earthquakes and Society (pp. 244–251)
EARTHQUAKE HAZARD
Write the letter of the correct answer in the space provided.

_____ **1.** What is a measurement of how likely an area is to have damaging earthquakes called?
a. gap hypothesis
b. seismic gap
c. earthquake hazard
d. earthquake frequency

EARTHQUAKE FORECASTING
Strength and Frequency

Use the terms from the following list to complete the sentences below.

seismic gap   frequency   gap hypothesis

2. The relationship between earthquake strength and _____

is based on the amount of energy released during earthquakes.

3. A method of forecasting where strong earthquakes are likely to occur in the

future is based on the _____.

4. An area along a fault where few earthquakes have occurred recently is called

a(n) _____.

Using the Gap Hypothesis
Write the letter of the correct answer in the space provided.

_____ **5.** The gap hypothesis may have helped predict which of the following earthquakes?
a. 1989 Loma Prieta earthquake
b. 1755 Lisbon earthquake
c. 1923 Tokyo earthquake
d. 1811–1812 New Madrid earthquake

REDUCING EARTHQUAKE DAMAGE

_____ **6.** What is the process of making older buildings more earthquake resistant called?
a. aftershocking   **c.** gapping
b. restructuring   **d.** retrofitting

Use the terms from the following list to complete the sentences below.

cross braces   flexible pipes   mass damper
base isolators   active tendon system

7. A weight placed in the roof of an earthquake-resistant building is called a(n)

_____.

8. A weight below an earthquake-resistant building that shifts to counteract the

building's movement is called the _____.

9. Earthquake-resistant buildings have _____ to prevent

waterlines and gas lines from breaking during earthquakes.

10. Earthquake-resistant buildings have _____ between floors

to counteract pressure during earthquakes.

11. During an earthquake, _____ act as shock absorbers to

prevent waves from traveling through the building.

ARE YOU PREPARED FOR AN EARTHQUAKE?

Before the Shaking Starts

Write the letter of the correct answer in the space provided.

_____ **12.** What should you do before an earthquake occurs?
    a. Safeguard your home.
    b. Be aware of aftershocks.
    c. Consume all your food and water.
    d. Put heavy objects on high shelves.

When the Shaking Starts

_____ **13.** Which of the following should you NOT do if you are indoors when
an earthquake strikes?
    a. Stay in the center of the room.
    b. Stay indoors.
    c. Go outside immediately.
    d. Crouch or lie under a table or desk.

_____ **14.** What should you do if you are outdoors when an earthquake strikes?
    a. Run back indoors.
    b. Stay outside, and lie face down.
    c. Climb a tree.
    d. Drive away in your car.

| Directed Reading A continued

After the Shaking Stops

_____ **15.** Which of the following is NOT an immediate danger after an earthquake?
  a. downed power lines
  b. nonperishable food
  c. broken glass
  d. fire hazards

_____ **16.** Which of the following should you NOT do after an earthquake?
  a. Stay out of damaged buildings.
  b. Return home if it is safe.
  c. Be aware that there may be aftershocks.
  d. Dispose of your first-aid kit.

TSUNAMIS

_____ **17.** What was the magnitude of the undersea earthquake that caused a tsunami in 2004?
  a. 5.5
  b. 6.4
  c. 9.3
  d. 4.2

Use the terms from the following list to complete the sentences below.

  Pacific Ocean   landslide   tsunami

18. A giant ocean wave caused by an earthquake is called a(n)

  _____ .

19. A natural event such as an undersea volcanic eruption, meteor impact, or

  _____ can cause a tsunami.

20. Tsunamis are monitored by most nations that border the

  _____ .

Name _____ Class _____ Date _____

Vocabulary and Section Summary A

What Are Earthquakes?

VOCABULARY

In your own words, write a definition of the following terms in the space provided.

1. earthquake

_____

_____

_____

2. elastic rebound

_____

_____

_____

3. seismic wave

_____

_____

_____

SECTION SUMMARY

Read the following section summary.

• Earthquakes are motions of the ground that happen as energy travels through rock.

• Earthquakes occur mainly near the edges of tectonic plates.

• Earthquakes are caused by elastic rebound, which is caused by sudden motions along faults. During elastic rebound, rock springs into its original shape and size as stress is released.

• Energy generated by earthquakes travels as body waves through Earth's interior or as surface waves along the surface of Earth.

Skills Worksheet

Vocabulary and Section Summary A

Earthquake Measurement
VOCABULARY
In your own words, write a definition of the following terms in the space provided.

1. epicenter

_____

_____

2. focus

_____

_____

3. magnitude

_____

_____

4. intensity

_____

_____

SECTION SUMMARY
Read the following section summary.

- An epicenter is the point on Earth's surface directly above where an earthquake started.
- The distance from a seismometer to an epicenter can be determined by using the lag time between P waves and S waves.
- An earthquake's epicenter can be located by triangulation.
- Magnitude is a measure of an earthquake's strength.
- Intensity is the effects of an earthquake.
- Important factors that determine the effects of an earthquake on a given area are magnitude, distance from the epicenter, local geology, and the type of construction.

Skills Worksheet

Vocabulary and Section Summary A

Earthquakes and Society

VOCABULARY

In your own words, write a definition of the following terms in the space provided.

1. seismic gap

_____

_____

_____

2. tsunami

_____

_____

_____

SECTION SUMMARY

Read the following section summary.

• Earthquakes and tsunamis can affect human societies.

• Earthquake hazard is a measure of how likely an area is to have earthquakes in the future.

• Scientists use their knowledge of the relationship between earthquake strength and frequency and of the gap hypothesis to forecast earthquakes.

• Homes, buildings, and bridges can be strengthened to decrease earthquake damage.

• People who live in earthquake zones should safeguard their homes against earthquakes and have an earthquake emergency plan.

• Tsunamis are giant ocean waves that may be caused by earthquakes on the sea floor.

Directed Reading A

Section: Why Volcanoes Form (pp. 266–269)

Write the letter of the correct answer in the space provided.

_____ **1.** What is a vent in Earth's crust through which magma and gases pass called?
   a. a trench
   b. a molten rock
   c. a mantle chamber
   d. a volcano

WHERE VOLCANOES FORM

_____ **2.** Where are many volcanoes found?
   a. at Earth's core
   b. in trenches
   c. at or near tectonic plate boundaries
   d. in dry climates

_____ **3.** Why are the plate boundaries surrounding the Pacific Ocean called the Ring of Fire?
   a. There is a huge lava plateau there.
   b. There are many active volcanoes in the area.
   c. There is a large ring-shaped crater there.
   d. The shallow water is a golden color.

_____ **4.** What is the name of the molten rock produced under Earth's surface?
   a. ash
   b. magma
   c. lava
   d. pressure

_____ **5.** Why does magma rise to Earth's surface?
   a. Magma is heavier than rock.
   b. Magma is less dense than the rock around it.
   c. Magma is harder than the rock around it.
   d. Magma is tricky.

Directed Reading A *continued*

Divergent Boundaries

Match the correct description with the correct term. Write the letter in the space provided.

_____ **6.** place where two tectonic plates move away from each other

_____ **7.** mountain chain created by underwater lava

a. mid-ocean ridge

b. divergent boundary

Convergent Boundaries

Match the correct description with the correct term. Write the letter in the space provided.

_____ **8.** place where two tectonic plates collide

_____ **9.** the movement of one tectonic plate under another tectonic plate

_____ **10.** deep depression formed when one plate gets pushed under another plate

a. trench

b. convergent boundary

c. subduction

Hot Spots

Match the correct description with the correct term. Write the letter in the space provided.

_____ **11.** volcanically active places that are not located at tectonic plate boundaries

_____ **12.** columns of hot rock that rise through Earth

a. hot spots

b. mantle plumes

HOW MAGMA FORMS

Write the letter of the correct answer in the space provided.

_____ **13.** Where does magma form?
   a. in Earth's crust and mantle
   b. in Earth's coat and crust
   c. in Earth's oceans and rivers
   d. on Earth's surface

_____ **14.** What causes magma to form?
   a. putty-like rock
   b. changes in temperature and pressure
   c. atoms of the mantle
   d. low temperatures

| Directed Reading A *continued*

Increasing Temperature

_____ **15.** How do increases in Earth's temperature affect rock?
   a. The rock can freeze and form more ice.
   b. The rock can melt and form magma.
   c. The rock will move horizontally.
   d. The rock is not affected by temperature.

Decreasing Pressure

_____ **16.** What happens when the pressure on a rock decreases?
   a. The hot rock expands and melts.
   b. The hot rock shrinks and cools.
   c. The hot rock moves to the side.
   d. The hot rock sinks to the bottom.

Adding Fluids

_____ **17.** How does the addition of fluids to hot rock cause the formation of magma?
   a. The rock sinks.
   b. The rock floats.
   c. The rock begins to melt.
   d. The rock begins to cool.

Skills Worksheet)

Directed Reading A

Section: Types of Volcanoes (pp. 270–277)

Write the letter of the correct answer in the space provided.

_____ 1. What determines how different types of volcanoes are created?
   a. mid-ocean ridges
   b. the color of the lava
   c. the formation and composition of the magma
   d. sea-floor spreading

VOLCANOES AT DIVERGENT BOUNDARIES

_____ 2. What is a rift zone?
   a. an area of deep cracks that forms when two tectonic plates are pulling away from each other
   b. an area of deep cracks that forms when two tectonic plates are colliding
   c. a place where there are many hot springs
   d. an area of deep cracks that forms when one tectonic plate slides under another tectonic plate

_____ 3. What is magma that flows onto Earth's surface called?
   a. lava
   b. fissure
   c. ash
   d. rock

Lava at Divergent Boundaries

_____ 4. What term is used to describe dark-colored magma that is rich in iron and magnesium?
   a. felsic
   b. mid-ocean ridge
   c. oceanic lithosphere
   d. mafic

_____ 5. What type of eruption generally occurs with lava that is low in silica?
   a. pyroclastic flows
   b. hot debris, ash, and gas shooting into the air
   c. explosive
   d. nonexplosive

| Directed Reading A *continued*

Mid-Ocean Ridges

_____ **6.** What is the name of an underwater mountain chain created where two tectonic plates are separating?
   a. divergent boundary
   b. rift zone
   c. mid-ocean ridge
   d. hot spring

_____ **7.** What is sea-floor spreading?
   a. the process of creating new underwater coral reefs
   b. the process in which new sea floor forms as older sea floor is pulled apart
   c. the process of creating tidal waves
   d. the process in which new sea floor forms as older sea floor is pushed together

Fissure Eruptions

_____ **8.** What is happening to Iceland due to frequent nonexplosive lava eruptions through fissures?
   a. Iceland is sinking.
   b. Iceland is getting larger.
   c. Iceland is getting smaller.
   d. Iceland is getting warmer.

VOLCANOES AT HOT SPOTS

_____ **9.** How does a hot spot form?
   a. A tectonic plate moves over a mantle plume.
   b. Magma flows over the ocean floor.
   c. An explosive eruption occurs.
   d. A nonexplosive eruption occurs.

_____ **10.** What may form with continuous eruptions at a hot spot?
   a. a convergent boundary
   b. a divergent boundary
   c. a volcano
   d. a new hot spot

Lava at Hot Spots

_____ **11.** Where does lava at hot spots originate?
   a. from tidal waves
   b. in Earth's mantle
   c. in Earth's atmosphere
   d. in the oceanic lithosphere

| Directed Reading A *continued*

Shield Volcanoes

_____ **12.** Where do shield volcanoes usually form?
   a. at hot spots
   b. at convergent boundaries
   c. at divergent boundaries
   d. at mid-ocean ridges

_____ **13.** How do shield volcanoes form?
   a. from tidal waves
   b. from continuous explosive eruptions
   c. from pyroclastic flows
   d. from continuous nonexplosive eruptions

Parts of a Volcano

_____ **14.** Where is a volcano's magma chamber?
   a. at the top of the vent
   b. deep underground
   c. on the volcano's slope
   d. outside the volcano

_____ **15.** What is a *vent*?
   a. a magma chamber
   b. an eruption
   c. an opening in Earth's crust
   d. an underground volcano

VOLCANOES AT CONVERGENT BOUNDARIES

_____ **16.** How does water affect the melting temperature of rock at a
   convergent boundary?
   a. Water lowers the melting temperature.
   b. Water raises the melting temperature.
   c. Water causes the rock to split apart.
   d. Water does not affect the melting temperature.

Lava at Convergent Boundaries

_____ **17.** What term is used to describe light-colored magma that is rich in
   silica and feldspars?
   a. felsic
   b. mid-ocean ridge
   c. oceanic lithosphere
   d. mafic

| Directed Reading A *continued*

_____ **18.** Which of the following statements is true?
    a. Silica-rich magma has a thin, runny consistency.
    b. Silica-rich magma allows gases to escape easily.
    c. Silica-rich magma causes explosive eruptions.
    d. Silica-rich magma is rarely associated with explosive eruptions.

Types of Pyroclastic Material

_____ **19.** What is pyroclastic material?
    a. molten rock
    b. magma that blasts into the air and hardens
    c. lava that flows underwater
    d. magma that remains underground too long

Pyroclastic Flows

_____ **20.** What is a very dangerous type of volcanic flow called?
    a. rift zone
    b. mid-ocean ridge
    c. oceanic lithosphere
    d. pyroclastic flow

Cinder Cone Volcanoes

Match the correct description with the correct term. Write the letter in the space provided.

_____ **21.** volcanoes made entirely of pyroclastic material

_____ **22.** volcanoes formed from layers of lava and pyroclastic materials

_____ **23.** another name for composite volcanoes

a. composite volcanoes

b. cinder cone volcanoes

c. stratovolcanoes

Directed Reading A

Section: Effects of Volcanic Eruptions (pp. 278–281)

Write the letter of the correct answer in the space provided.

_____ **1.** What can a volcanic explosion affect globally?
    a. ocean salinity
    b. rising water levels of lakes
    c. climate
    d. traffic

NEGATIVE IMPACTS OF VOLCANIC ERUPTIONS

_____ **2.** Which of the following describes a climate change caused by a volcanic eruption?
    a. Temperatures rise.
    b. Ash blocks sunlight, causing temperatures to drop.
    c. Burned land creates dry conditions.
    d. Volcanic eruptions don't cause climate changes.

_____ **3.** Which of the following describes a global impact from the temperature change caused by a volcanic eruption?
    a. crop failure
    b. less drinking water
    c. homes collapsing
    d. farm equipment failure

Local Effects of Volcanic Eruptions

_____ **4.** What type of eruption is most destructive?
    a. nonexplosive
    b. underwater
    c. explosive
    d. atmospheric

_____ **5.** What are lahars?
    a. tidal waves caused by underwater eruptions
    b. mudflows made from volcanic ash and water
    c. large vents in Earth's crust
    d. cooled rocks from a volcanic eruption

_____ **6.** What is a specific health problem that volcanic ash can cause?
    a. measles
    b. heart disease
    c. muscle loss
    d. respiratory problems

Global Effects of Volcanic Eruptions

_____ **7.** What can happen to Earth's temperature for several years after a large volcanic eruption?
  a. The global temperature may increase.
  b. The global temperature may decrease.
  c. Global warming may increase.
  d. Earth's temperature is not affected by volcanic activity.

BENEFITS OF VOLCANIC ERUPTIONS

_____ **8.** What are two benefits provided by volcanoes?
  a. hot lava and ash
  b. warmer temperatures and deeper oceans
  c. fresh drinking water and new medicines
  d. fertile soil and construction materials

Volcanic Soils

_____ **9.** Why is volcanic soil so fertile?
  a. It contains toxins.
  b. It contains many nutrients.
  c. It contains excessive amounts of water.
  d. It contains heat from volcanic rocks.

Geothermal Energy

_____ **10.** What is the heated water located in volcanic rocks called?
  a. drinking water
  b. acid rain
  c. geothermal water
  d. volcanic flow

_____ **11.** What is one use of geothermal energy?
  a. to generate electricity
  b. to eliminate traffic jams
  c. to create habitats for animals
  d. to bury crops

_____ **12.** In Reykjavik, Iceland, what percentage of homes are heated by geothermal power?
  a. 50%
  b. 75%
  c. 85%
  d. 100%

Other Benefits of Volcanic Eruptions

_____ **13.** What is one way basalt and pumice are used today?
   a. in the production of concrete
   b. to grow food
   c. in the production of hot springs
   d. to reduce the effects of tidal waves

_____ **14.** Why is pumice added to soaps and cleaners?
   a. because it is abrasive
   b. because it is smooth
   c. because it absorbs moisture
   d. because it allows air and water to circulate through soil

Vocabulary and Section Summary A

Why Volcanoes Form
VOCABULARY
In your own words, write a definition of the following terms in the space provided.

1. volcano

_____

_____

_____

2. magma

_____

_____

_____

SECTION SUMMARY
Read the following section summary.

- A volcano is a vent or fissure in Earth's surface through which magma and gases pass.

- Most volcanoes are located at tectonic plate boundaries.

- Volcanic activity occurs at divergent plate boundaries, convergent plate boundaries, and hot spots.

- Magma forms when the temperature of a rock increases, when the pressure on a rock decreases, or when water lowers the melting temperature of a rock.

Skills Worksheet

Vocabulary and Section Summary A

Types of Volcanoes

VOCABULARY

In your own words, write a definition of the following terms in the space provided.

1. lava

_____

_____

_____

2. mafic

_____

_____

_____

3. felsic

_____

_____

_____

SECTION SUMMARY

Read the following section summary.

- Mafic lava erupts quietly through cracks, or fissures, in the lithosphere at divergent boundaries.

- At hot spots, continuous eruptions of mafic magma form chains of volcanoes above mantle plumes.

- Shield volcanoes form from the mafic lava erupted at hot spots.

- At convergent boundaries, eruptions of silica-rich magma are often explosive.

- Composite volcanoes form from the felsic lava erupted at convergent boundaries.

Vocabulary and Section Summary A

Effects of Volcanic Eruptions

VOCABULARY

In your own words, write a definition of the following term in the space provided.

1. lahar

_____

_____

_____

SECTION SUMMARY

Read the following section summary.

• Volcanic eruptions can have local effects on humans and on wildlife habitats.

• When ash and gases from a large volcanic eruption spread around the planet, they may absorb and scatter enough sunlight to cause a temporary decrease in the average global temperature.

• Benefits that volcanoes provide to humans and to the environment include fertile soil, a renewable energy source, and construction materials.

Directed Reading A

Section: Weathering (pp. 298–303)

Write the letter of the correct answer in the space provided.

_____ 1. What is the process by which rocks are broken down physically or chemically called?
   a. abrasion
   b. frost action
   c. weathering
   d. ice wedging

MECHANICAL WEATHERING

_____ 2. What is it called when rocks are broken into smaller pieces by physical means?
   a. mechanical weathering
   b. acid oxidation
   c. chemical weathering
   d. acid precipitation

_____ 3. Which of the following is an agent of mechanical weathering?
   a. sunlight
   b. oxidation
   c. weak acids
   d. gravity

Ice Wedging

_____ 4. What is the alternate freezing and thawing of soil and rock called?
   a. frost action
   b. ice weathering
   c. physical abrasion
   d. chemical weathering

_____ 5. What type of frost action occurs when a water-filled crack in a rock widens from the freezing and thawing of the water?
   a. chemical weathering
   b. ice wedging
   c. oxidation
   d. abrasion

Abrasion

_____ **6.** What is the grinding and wearing away of rock surfaces by other rocks or sand called?
   a. precipitation
   b. chemical weathering
   c. abrasion
   d. oxidation

_____ **7.** What causes abrasion of rocks and pebbles in fast-flowing streams?
   a. gravity
   b. water
   c. acid precipitation
   d. wind

_____ **8.** What causes abrasion when one rock slides downhill and grinds against another rock?
   a. gravity
   b. water
   c. acid precipitation
   d. wind

Exfoliation

_____ **9.** How does erosion of a rock surface cause expansion of the underlying rock?
   a. by decreasing pressure on it
   b. by increasing pressure on it
   c. by breaking the rock
   d. by smoothing the rock

_____ **10.** What process causes sheets of rock to peel away from the underlying rock?
   a. abrasion
   b. exfoliation
   c. ice wedging
   d. acid precipitation

Plant Growth

_____ **11.** Which is an example of a plant causing mechanical weathering of a rock?
   a. lichen acid breaking a rock
   b. gophers burrowing in the ground
   c. tree roots cracking a rock
   d. strong rain washing away rocks

Animal Actions

_____ **12.** How can animals that live in soil cause mechanical weathering?
   a. by causing oxidation
   b. by drinking water
   c. by eating plants
   d. by burrowing

CHEMICAL WEATHERING

_____ **13.** What is it called when rocks break down due to chemical reactions?
   a. ice wedging
   b. mechanical weathering
   c. burrowing
   d. chemical weathering

_____ **14.** Which of the following is a common agent of chemical weathering?
   a. sunlight
   b. acids
   c. gravity
   d. abrasion

How Water Chemically Breaks Down Rock

_____ **15.** What kind of weathering takes place when a sugar cube dissolves in water?
   a. differential weathering
   b. chemical weathering
   c. mechanical weathering
   d. desertification

Acid Precipitation

_____ **16.** What is rain, sleet, or snow that contains a lot of acid called?
   a. chemical weathering
   b. acid evaporation
   c. acid precipitation
   d. fossil fuel

_____ **17.** Which of the following can help to cause acid precipitation?
   a. burning fossil fuels
   b. rusting iron
   c. burrowing animals
   d. falling rocks

| Directed Reading A *continued*

How Acids in Groundwater Weather Rock

_____ **18.** What can result from acids in groundwater coming into contact with limestone?
   a. lichens
   b. cavern formation
   c. oxidation
   d. acid precipitation

How Acids in Living Things Weather Rock

_____ **19.** What products from living things can slowly break down the rocks that organisms may touch?
   a. neutral acids
   b. dilute acids
   c. strong acids
   d. acid precipitates

How Air Chemically Weathers Rock

_____ **20.** What is it called when an element combines with oxygen to form an oxide?
   a. abrasion
   b. mechanical weathering
   c. acid precipitation
   d. oxidation

_____ **21.** When oxygen in the air reacts with iron, what substance is formed?
   a. abrasion
   b. mechanical weathering
   c. acid precipitation
   d. rust

Skills Worksheet

Directed Reading A

Section: Rates of Weathering (pp. 304–307)
Write the letter of the correct answer in the space provided.

_____ 1. Which answer best describes weathering?
    a. a slow process
    b. a result of rust
    c. a result of ice wedging
    d. a fast process

DIFFERENTIAL WEATHERING

_____ 2. What is the process by which softer, less weather resistant rocks wear away at a faster rate than harder, more weather resistant rocks?
    a. abrasion
    b. differential weathering
    c. oxidation
    d. acidic weathering

THE SURFACE AREA OF ROCKS

_____ 3. On what part of a rock does weathering take place?
    a. the outer surface
    b. throughout the rock
    c. the inner core
    d. between core and surface

_____ 4. What happens to the rate of weathering if a big rock is broken into smaller pieces?
    a. The rate decreases.
    b. The rate is unchanged.
    c. The rate increases.
    d. The rate stops.

WEATHERING AND CLIMATE

_____ 5. What do we call the average weather condition in an area over a long period of time?
    a. weather
    b. humidity
    c. temperature
    d. climate

❘ Directed Reading A *continued*

Temperature

_____ **6.** In what type of climate does chemical weathering happen most quickly?
   a. warm and humid
   b. hot and dry
   c. cold and damp
   d. cold and dry

Moisture

_____ **7.** What kind of weathering causes a mailbox to rust?
   a. ice wedging
   b. chemical
   c. mechanical
   d. thawing

OTHER FACTORS THAT AFFECT WEATHERING

_____ **8.** How do rocks weather at higher elevations?
   a. They have a lower rate of weathering than rocks at lower elevations.
   b. They have a higher rate of weathering than rocks at lower elevations.
   c. They are not affected by weathering.
   d. They are affected only by chemical weathering.

_____ **9.** How does the slope of the ground affect its rate of weathering?
   a. Steeper ground weathers more slowly.
   b. Slope does not affect weathering.
   c. Steeper ground weathers more quickly.
   d. Less steep ground never weathers.

_____ **10.** How can organisms help cause weathering?
   a. They increase stream flow.
   b. They add air pollution.
   c. They produce acid.
   d. They decrease rock volume.

Directed Reading A

Section: From Bedrock to Soil (pp. 308–313)
THE SOURCE OF SOIL

Match the correct description with the correct term. Write the letter in the space provided.

_____ **1.** soil that is carried away from its parent rock by wind, water, ice, or gravity.

_____ **2.** the layer of rock beneath the soil

_____ **3.** rock that breaks down into mineral fragments that form a soil

_____ **4.** a loose mixture of mineral fragments, organic material, water, and air that can support the growth of vegetation

a. soil

b. transported soil

c. parent rock

d. bedrock

SOIL PROPERTIES
Soil Composition

Write the letter of the correct answer in the space provided.

_____ **5.** What does soil contain that affects its ability to support plant life?
  a. air mixture
  b. mineral particles
  c. organic material
  d. moisture

Soil Texture

_____ **6.** What do the relative amounts of different-sized soil particles determine?
  a. soil structure
  b. soil texture
  c. soil horizons
  d. soil fertility

_____ **7.** What does soil texture affect?
  a. consistency and transportation
  b. consistency and infiltration
  c. transportation and infiltration
  d. transportation and sedimentation

❙ Directed Reading A *continued*

Soil Fertility

_____ **8.** What is soil's ability to hold nutrients and to supply nutrients to a plant?
 a. soil structure
 b. soil texture
 c. soil horizons
 d. soil fertility

_____ **9.** What provides soil nutrients from the decayed remains of plants and animals?
 a. parent rock
 b. infiltration
 c. air flow
 d. humus

Soil Horizons

_____ **10.** What do we call the layers that are created because of the way soil forms?
 a. parent rock
 b. sediment
 c. topsoil
 d. soil horizons

_____ **11.** What do we call the layer of soil that contains the most humus?
 a. horizon
 b. parent rock
 c. topsoil
 d. bedrock

Soil pH

_____ **12.** What is soil that has a pH below 7 called?
 a. neutral
 b. acidic
 c. basic
 d. midpoint

_____ **13.** How can soil pH affect plant growth?
 a. by blocking soil nutrients
 b. by making them acidic
 c. by making them basic
 d. by filtering sunlight

_____ **14.** In what pH range do most plants grow best?
  a. 7.5 to 8.5
  b. 5.5 to 7
  c. 4 to 5
  d. at no range

_____ **15.** What partly determines soil pH?
  a. soil horizons
  b. hungry plants
  c. humus
  d. parent rock

SOIL AND CLIMATE
Tropical Climates

_____ **16.** Which of the following is true of soil in tropical rain forests?
  a. It has little plant life.
  b. It is nutrient poor.
  c. It gets little water.
  d. It has very little humus.

Desert and Arctic Climates

_____ **17.** Which of the following is a characteristic of desert and arctic climates?
  a. low rates of chemical weathering
  b. a high rate of weathering
  c. extensive annual rainfall
  d. low removal of soil nutrients

Temperate Forest and Grassland Climates

_____ **18.** Which of the following is true of temperate forest and grassland climates?
  a. Little weathering occurs.
  b. Temperatures are steady.
  c. Productive soils develop.
  d. Soil nutrients are lost.

Skills Worksheet

Directed Reading A

Section: Soil Conservation (pp. 314–317)

Write the letter of the correct answer in the space provided.

_____ 1. What do we call a resource, such as soil, that is difficult to replace?
  a. recoverable
  b. nonrecoverable
  c. nonrenewable
  d. renewable

_____ 2. What do we call methods to keep soil fertile by protecting it from erosion and nutrient loss?
  a. soil degradation
  b. soil terracing
  c. soil desertification
  d. soil conservation

THE IMPORTANCE OF SOIL

_____ 3. What will happen to animals if the plants they depend on fail to get enough nutrients from the soil?
  a. They will fail to get plant nutrients.
  b. They will get plant nutrients.
  c. They will fail to get water nutrients.
  d. They will get water nutrients.

Housing

_____ 4. What is the region where a plant or animal lives called?
  a. soil
  b. habitat
  c. contour
  d. cover

Water Storage

_____ 5. If soil did not hold water, what would be the result?
  a. not enough ice wedging
  b. not enough soil conservation
  c. not enough moisture for plant growth
  d. not enough acid precipitation

SOIL DAMAGE AND LOSS

_____ **6.** What is the process when plants cannot hold or help cycle water called?
    a. desertification
    b. habitation
    c. erosion
    d. terracing

Soil Erosion

_____ **7.** What is the transportation of sediment by wind, water, ice, or gravity from one location to another called?
    a. degradation
    b. habitation
    c. erosion
    d. terracing

SOIL CONSERVATION ON FARMLAND

Match the correct description with the correct term. Write the letter in the space provided.

_____ **8.** when farmers plow across the slope of a hill to prevent erosion

_____ **9.** when farmers change a steep field into a series of smaller, flatter fields

_____ **10.** the practice of leaving old stalks to provide cover from the rain

    a. contour plowing
    b. no-till farming
    c. terracing

Cover Crops and Crop Rotation

Write the letter of the correct answer in the space provided.

_____ **11.** What are plants such as soybeans and peanuts that restore nutrients to the soil called?
    a. rotating crops
    b. root crops
    c. contour crops
    d. cover crops

_____ **12.** What process slows nutrient depletion by planting different crops that use different soil nutrients?
    a. crop rotation
    b. root cropping
    c. crop contouring
    d. cover cropping

Name _____ Class _____ Date _____

Vocabulary and Section Summary A

Weathering
VOCABULARY

In your own words, write a definition of the following terms in the space provided.

1. weathering

_____

_____

2. mechanical weathering

_____

_____

3. abrasion

_____

_____

4. exfoliation

_____

_____

5. chemical weathering

_____

_____

6. acid precipitation

_____

_____

SECTION SUMMARY

Read the following section summary.

- Ice wedging is a form of mechanical weathering in which water seeps into cracks in rock and then freezes and expands.

- Wind, water, and gravity cause mechanical weathering by abrasion.

- Animals and plants cause mechanical weathering by mixing the soil and breaking apart rocks.

- Water, acids, and oxygen in the air chemically weather rock by reacting chemically with elements in the rock.

- Oxidation is the process by which oxygen from the air reacts with iron in rocks.

Skills Worksheet

Vocabulary and Section Summary A

Rates of Weathering

VOCABULARY

In your own words, write a definition of the following term in the space provided.

1. differential weathering

_____

_____

SECTION SUMMARY

Read the following section summary.

- Hard rocks weather more slowly than soft rocks.
- The larger the surface area–to-volume ratio of a rock is, the faster the rock will wear down.
- Chemical weathering occurs faster in warm, wet climates than in hot, dry climates.
- Rates of weathering are affected by elevation, by the slope of the ground, and by living things.

Skills Worksheet

Vocabulary and Section Summary A

From Bedrock to Soil

VOCABULARY

In your own words, write a definition of the following terms in the space provided.

1. soil

_____

_____

2. parent rock

_____

_____

3. bedrock

_____

_____

4. humus

_____

_____

SECTION SUMMARY

Read the following section summary.

• Soil forms from the weathering of bedrock.

• Soil texture affects how soil can be worked for farming and how well water passes through soil.

• The ability of soil to provide nutrients so that plants can survive and grow is called *soil fertility*.

• The pH of a soil influences which nutrients plants can take up from the soil.

• Different climates have different types of soil, depending on the temperature and rainfall.

• The characteristics of soil affect the number and types of organisms that an area can support.

Vocabulary and Section Summary A

Soil Conservation
VOCABULARY
In your own words, write a definition of the following terms in the space provided.

1. soil conservation

_____

_____

2. erosion

_____

_____

_____

SECTION SUMMARY
Read the following section summary.

• Soil forms slowly over hundreds or thousands of years. Therefore, soil is considered a nonrenewable resource.

• Soil is important because plants grow in soil, animals live in soil, and water is stored in soil.

• Soil can be eroded by water running downhill or by wind.

• Soil erosion and soil damage can be prevented by no-till farming, contour plowing, terracing, using cover crops, and practicing crop rotation.

Skills Worksheet

Directed Reading A

Section: Shoreline Erosion and Deposition (pp. 332–339)

Use the terms from the following list to complete the sentences below.

shoreline   sand

1. When waves crash into rock over long periods of time, the rock breaks down

   into smaller pieces called _____.

2. A body of water meets land at a place called a(n)

   _____.

WAVE ENERGY

Write the letter of the correct answer in the space provided.

_____ 3. What does the size of a wave depend on?
   a. sunshine
   b. sand
   c. water
   d. wind

Wave Trains

_____ 4. What are waves that travel in groups called?
   a. wave trains
   b. surf
   c. wave periods
   d. ocean waves

_____ 5. What is the length of time between breaking waves called?
   a. wave train
   b. surf
   c. wave period
   d. ripples

_____ 6. What are breaking waves called?
   a. wave trains
   b. surf
   c. wave period
   d. ocean waves

| Directed Reading A *continued*

The Pounding Surf

_____ **7.** What do breaking waves release?
   a. sand
   b. energy
   c. wave trains
   d. water

_____ **8.** What do waves make when they break rock into smaller pieces?
   a. rocks
   b. pebbles
   c. sand
   d. boulders

WAVE EROSION

Match the correct definition with the correct term. Write the letter in the space provided.

_____ **9.** have steep slopes formed when waves undercut rock

_____ **10.** columns of rock that were once part of the mainland

_____ **11.** large holes cut in rocks at the base of sea cliffs

_____ **12.** form when waves erode sea caves

a. sea caves

b. sea stacks

c. sea arches

d. sea cliffs

Shaping a Shoreline

Write the letter of the correct answer in the space provided.

_____ **13.** What happens when large waves from storms hit the shoreline?
   a. Huge chunks of rock are removed.
   b. Huge chunks of rock are deposited.
   c. Only very small pieces of rock are removed.
   d. Erosion does not takes place.

WAVE DEPOSITS

_____ **14.** What do you call the part of the shoreline made up of materials deposited by waves?
   a. headland
   b. wave-cut terrace
   c. barrier spit
   d. beach

| Directed Reading A *continued*

Beach Materials

_____ **15.** What can beaches be made of where stormy seas are common?
   a. pebbles and boulders
   b. coral
   c. eroded lava
   d. quartz sand

Beach Composition

_____ **16.** What are most beaches made of?
   a. sand
   b. shells
   c. coral
   d. lava

_____ **17.** Where are black sand beaches found?
   a. California
   b. Hawaii
   c. Florida
   d. the Virgin Islands

California Beaches

_____ **18.** Which of the following statements is NOT true about California beaches?
   a. California has both rocky and sandy beaches.
   b. The mineral composition of California beaches varies.
   c. The beach at Carmel is made of quartz and feldspar.
   d. All California beaches are made of finely ground coral.

Beach Size

_____ **19.** What can happen to California beaches in winter?
   a. They can become wider.
   b. They can become narrower.
   c. They can become rockier.
   d. They can become a different color.

Shore Currents

_____ **20.** What is the name of a current near the shore that pulls objects out to sea?
   a. wave
   b. barrier spit
   c. undertow
   d. longshore current

Longshore Currents

_____ **21.** What pattern does a longshore current make?
    a. up and down
    b. zigzag
    c. circle
    d. square

Offshore Deposits

_____ **22.** What do you call a ridge of sand, gravel, or shells in the water?
    a. barrier spit
    b. sandbar
    c. current
    d. shoreline

CALIFORNIA ISLANDS

_____ **23.** When did California's Channel Islands form?
    a. millions of years ago
    b. a thousand years ago
    c. a hundred years ago
    d. twenty years ago

Skills Worksheet

Directed Reading A

Section: Wind Erosion and Deposition (pp. 340–343)
WIND EROSION
Write the letter of the correct answer in the space provided.

_____ 1. What helps keep wind from blowing sand and soil away?
    a. water
    b. rock
    c. plants
    d. deserts

Match the correct definition with the correct term. Write the letter in the space provided.

_____ 2. Strong winds cause sand-sized particles to skip and bounce.

_____ 3. Rock is ground and worn down by other rock or by sand.

_____ 4. Wind creates a surface that is made of pebbles and broken rocks.

_____ 5. Wind blows away fine, dry soil particles.

    a. saltation
    b. desert pavement
    c. abrasion
    d. deflation

WIND-DEPOSITED MATERIALS
Write the letter of the correct answer in the space provided.

_____ 6. What determines the amount and size of the particles that wind can carry?
    a. wind speed
    b. wind direction
    c. deflation
    d. abrasion

Dunes
Match the correct definition with the correct term. Write the letter in the space provided.

_____ 7. obstacles that cause wind to slow down

_____ 8. a mound of wind-deposited sand

_____ 9. a place where dunes are common

    a. dune
    b. deserts
    c. plants

| Directed Reading A *continued*

California Dunes
Write the letter of the correct answer in the space provided.

_____ **10.** What is the largest desert and dune system in California?
- **a.** Monterey Bay Dunes
- **b.** Algodones Sand Dunes
- **c.** the Painted Desert
- **d.** the Channel Islands

The Movement of Dunes

_____ **11.** Which way do sand dunes move?
- **a.** uphill
- **b.** downhill
- **c.** away from the direction of the wind
- **d.** in the direction of the wind

_____ **12.** What is the steeply sloped side of a dune called?
- **a.** slip face
- **b.** deflation hollow
- **c.** windward slope
- **d.** desert pavement

Skills Worksheet

Directed Reading A

Section: Erosion and Deposition by Ice (pp. 344–347)

Write the letter of the correct answer in the space provided.

_____ 1. What is the name for a large mass of moving ice?
    a. glacier
    b. till
    c. moraine
    d. drift

_____ 2. What is the name for glaciers that form in mountainous areas?
    a. alpine glaciers
    b. continental glaciers
    c. mountain glaciers
    d. valley glaciers

_____ 3. What is a glacier that spreads across an entire continent called?
    a. alpine glacier
    b. continental glacier
    c. mountain glacier
    d. valley glaciers

GLACIERS—RIVERS OF ICE
How Glaciers Move

_____ 4. What are two ways glaciers move?
    a. by melting and drying
    b. by sliding and flowing
    c. by melting and erosion
    d. by widening and straightening

LANDFORMS CREATED BY GLACIERS

Match the correct definition with the correct term. Write the letter in the space provided.

_____ 5. bowl-shaped depressions

_____ 6. sharp, pyramid-shaped peaks

_____ 7. jagged ridges

_____ 8. formed by a glacier eroding a river valley

    a. cirques
    b. arêtes
    c. horns
    d. U-shaped valleys

Directed Reading A *continued*

TYPES OF GLACIAL DEPOSITS

Write the letter of the correct answer in the space provided.

_____ **9.** What is the name of the material that is carried and deposited by glaciers?
- **a.** glacial drift
- **b.** outwash plain
- **c.** till
- **d.** rock

Till Deposits

_____ **10.** What is rock material that is deposited directly by melting ice called?
- **a.** till
- **b.** glacial drift
- **c.** valley
- **d.** moraine

_____ **11.** What are the most common till deposits?
- **a.** horns
- **b.** sediment
- **c.** valleys
- **d.** moraines

Match the correct definition with the correct term. Write the letter in the space provided.

_____ **12.** formed along the sides of glaciers

_____ **13.** formed when valley glaciers with lateral moraines meet

_____ **14.** formed from unsorted materials left beneath a glacier

_____ **15.** formed when sediment drops at the front of a glacier

- **a.** ground moraine
- **b.** terminal moraine
- **c.** lateral moraine
- **d.** medial moraine

Stratified Drift

Use the terms from the following list to complete the sentences below.

    outwash plain   kettle   stratified drift

16. A glacial deposit that has been sorted and layered is a(n)

_____.

17. Material spread out over a big area in front of a glacier is

a(n) _____.

18. A depression that forms a lake is a(n) _____.

Skills Worksheet

Directed Reading A

Section: Erosion and Deposition by Mass Movement (pp. 348–351)

Write the letter of the correct answer in the space provided.

_____ **1.** What makes rocks and soil move downslope?
   a. gravity
   b. deposition
   c. wind
   d. mass movement

_____ **2.** What is the name for the movement of any material downslope?
   a. mass movement
   b. deposition
   c. angle of repose
   d. loose material

ANGLE OF REPOSE

_____ **3.** What is the steepest angle at which loose material will not move downslope?
   a. mass movement
   b. deposition
   c. angle of repose
   d. angle of movement

_____ **4.** Which of the following is NOT a factor that decides the angle of repose?
   a. the shape of the material
   b. the weight of the material
   c. the color of the material
   d. the moisture level of the material

RAPID MASS MOVEMENT
Rock Falls

Match the correct definition with the correct term. Write the letter in the space provided.

_____ **5.** loose rocks that fall down a steep slope

_____ **6.** sudden movement of rock and soil down a slope

_____ **7.** flow of mud and water

_____ **8.** slow downhill movement of rock

a. rock fall
b. creep
c. landslide
d. mudflow

| Directed Reading A *continued*

Landslides

Write the letter of the correct answer in the space provided.

_____ **9.** Which of the following is NOT true about a landslide?

   a. A landslide can bury wildlife habitats.

   b. A landslide is a sudden movement of rock and soil.

   c. A landslide can carry away plants and animals.

   d. A landslide is never destructive.

Mudflows

_____ **10.** Where do mudflows commonly happen?

   a. on beaches

   b. in mountainous regions

   c. in the ocean

   d. in flat regions

Creep

_____ **11.** What is the very, very slow movement of material downhill called?

   a. landslide

   b. mudflow

   c. creep

   d. slump

_____ **12.** Which of the following does NOT contribute to creep?

   a. water

   b. plant roots

   c. sunlight

   d. burrowing animals

MASS MOVEMENT AND LAND USE

_____ **13.** Which of the following is a beneficial long-term effect of mass movement?

   a. Fresh minerals are exposed.

   b. Property is damaged.

   c. Wildlife habitats are destroyed.

   d. Plants and animals are buried.

Vocabulary and Section Summary A

Shoreline Erosion and Deposition

VOCABULARY

In your own words, write a definition of the following terms in the space provided.

1. shoreline

_____

_____

2. beach

_____

_____

3. undertow

_____

_____

4. longshore current

_____

_____

SECTION SUMMARY

Read the following section summary.

• A wave is a disturbance in the water that can be caused by wind.

• As waves break against a shoreline, their energy breaks rocks down into sand.

• Six shoreline features that are created by wave erosion are sea cliffs, sea stacks, sea caves, sea arches, headlands, and wave-cut terraces.

• Beaches are made from material deposited by rivers and waves.

• California beaches can be rocky or sandy and can have different mineral compositions.

• Longshore currents cause sand to move in a zigzag pattern along the shore.

• Longshore currents can deposit eroded sediment offshore.

Skills Worksheet

Vocabulary and Section Summary A

Wind Erosion and Deposition

VOCABULARY

In your own words, write a definition of the following terms in the space provided.

1. saltation

_____

_____

2. deflation

_____

_____

3. abrasion

_____

_____

4. dune

_____

_____

SECTION SUMMARY

Read the following section summary.

- Areas that have little plant cover and desert areas that are covered with fine rock material are more vulnerable to wind erosion than other areas are.

- Saltation is the process in which sand-sized particles move in the direction of the wind.

- Desert pavement, deflation hollows, and dunes are landforms that are created by wind erosion and deposition.

- Dunes move in the direction that the wind blows.

Vocabulary and Section Summary A

Erosion and Deposition by Ice

VOCABULARY

In your own words, write a definition of the following terms in the space provided.

1. glacier

_____

_____

2. glacial drift

_____

_____

3. till

_____

_____

4. stratified drift

_____

_____

SECTION SUMMARY

Read the following section summary.

• Glaciers are masses of moving ice that shape the landscape by eroding and depositing material.

• Glaciers move by sliding or by flowing.

• Alpine glaciers can carve cirques, arêtes, horns, U-shaped valleys, and hanging valleys.

• Two types of glacial deposits are till and stratified drift.

• Deposition of sediment by glaciers can form several landforms, including kettles.

Skills Worksheet

Vocabulary and Section Summary A

Erosion and Deposition by Mass Movement
VOCABULARY
In your own words, write a definition of the following terms in the space provided.

1. mass movement

_____

_____

2. rock fall

_____

_____

3. landslide

_____

_____

4. mudflow

_____

_____

5. creep

_____

_____

SECTION SUMMARY
Read the following section summary.

• Gravity causes rocks and soil to move downslope.

• If the slope on which material rests is greater than the angle of repose, mass movement will occur.

• Four types of mass movement are rock falls, landslides, mudflows, and creep.

• Landslides may destroy buildings and change wildlife habitats.

Directed Reading A

Section: The Active River (pp. 366–373)
RIVERS: AGENTS OF EROSION
Write the letter of the correct answer in the space provided.

_____ **1.** What process moves soil and sediment from one location to another?
   a. water cycle
   b. erosion
   c. deposition
   d. pollution

_____ **2.** What is a river that shapes Earth's landscape an agent of?
   a. evaporation
   b. condensation
   c. gravitation
   d. erosion

THE WATER CYCLE

_____ **3.** What is the continuous movement of water between the atmosphere, the land, and the oceans?
   a. water cycle
   b. erosion
   c. alluvial fans
   d. levees

_____ **4.** What is the source of energy behind the water cycle?
   a. the sun
   b. the moon
   c. Earth's gravity
   d. soil erosion

| Directed Reading A *continued*

Match the correct description with the correct term. Write the letter in the space provided.

_____ **5.** water from oceans and Earth's surface that changes to water vapor

_____ **6.** water that flows over land into streams and rivers

_____ **7.** rain, snow, sleet, or hail that falls from clouds

_____ **8.** water vapor that cools and changes into water droplets that form clouds

_____ **9.** water that moves downward through spaces in soil

a. evaporation

b. percolation

c. condensation

d. precipitation

e. runoff

RIVER SYSTEMS

Write the letter of the correct answer in the space provided.

_____ **10.** What is a network of streams and rivers that drains an area of its runoff?
    a. river system
    b. alluvial fan
    c. watershed
    d. floodplain

_____ **11.** What is a stream that flows into a lake or larger stream called?
    a. delta
    b. tributary
    c. network
    d. river system

Watersheds and Divides

_____ **12.** What is the area drained by a river system called?
    a. downflow
    b. overflow
    c. watershed
    d. floodplain

_____ **13.** What is the area of higher ground that separates two watersheds called?
    a. divide
    b. incline
    c. decline
    d. runoff

Directed Reading A *continued*

STREAM EROSION

_____ **14.** What is the path that a stream follows?
    a. tributary
    b. watershed
    c. channel
    d. divide

_____ **15.** What is the measure of change in elevation over a distance called?
    a. altitude
    b. gradient
    c. velocity
    d. tributary

_____ **16.** A river or stream that has a high gradient has
    a. a lot of energy for erosion.
    b. no energy for erosion.
    c. an unchanged amount of energy for erosion.
    d. very little energy for erosion.

_____ **17.** What is the amount of water a stream or river carries in a given time?
    a. gradient
    b. divide
    c. erosion
    d. discharge

_____ **18.** If a stream's discharge increases, what does its rate of erosion do?
    a. It decreases.
    b. It increases.
    c. It is unchanged.
    d. It is unclear.

_____ **19.** What are the materials carried by a stream called?
    a. bedrock
    b. suspension
    c. load
    d. solution

Use the terms from the following list to complete the sentences below.

bed load   dissolved load
erosion   suspended load

20. A load that is carried dissolved in water is a(n) _____.

21. A load of pebbles and boulders bounced along a stream bed is a(n)

_____.

22. A stream with a load of large particles has a high rate of

_____.

23. A muddy-looking load of soil and small rocks is called a(n)

_____.

DESCRIBING RIVERS

Use the terms from the following list to complete the sentences below.

rejuvenated river   terraces   old river
youthful river   mature river

24. A river with a channel deeper than it is wide is a(n) _____.

25. A river with a channel wider than it is deep is a(n) _____.

26. A river with wide, flat floodplains is a(n) _____.

27. A river where land is raised by tectonic activity is a(n)

_____.

28. Steplike formations on both sides of a stream valley are called

_____.

Skills Worksheet

Directed Reading A

Section: Stream and River Deposits (pp. 374–377)
Write the letter of the correct answer in the space provided.

_____ 1. Which body of water erodes and moves soil and rock?
    a. lake
    b. spring
    c. river
    d. aquifer

_____ 2. How can erosion by rivers be helpful?
    a. It renews soils.
    b. It causes precipitation.
    c. It cleans rivers.
    d. It changes wind currents.

STREAM DEPOSITS

_____ 3. What is the process in which material carried by a river is laid down?
    a. deposition
    b. reposition
    c. meandering
    d. sedimentation

_____ 4. What are the rocks and soil that streams and rivers deposit?
    a. landforms
    b. sediment
    c. deltas
    d. currents

_____ 5. What is a fan-shaped deposit of material at the mouth of a stream?
    a. an alluvial fan
    b. a floodplain
    c. a delta
    d. a placer deposit

_____ 6. Why do deltas cause coastlines to grow?
    a. They are made from mud deposits, which form new land surfaces.
    b. They are made from rocks, which break incoming waves.
    c. They are made of plants, which take up more space on the coastline.
    d. They deposit water, which causes the ocean to grow larger.

_____ **7.** Which of the following helped to form the Mississippi Delta?
   a. River currents were delivered from upstream.
   b. Plant life grew in the delta area.
   c. Mud particles were delivered from upstream.
   d. Animals moved into the delta area.

_____ **8.** What is formed when a fast-moving stream flows onto a flat plain and slows down quickly?
   a. a deposition
   b. a floodplain
   c. an alluvial fan
   d. a delta

FLOODS

_____ **9.** When a stream floods, which of the following may happen?
   a. Its banks strengthen.
   b. Its path changes.
   c. Its channel deepens.
   d. Its discharge lessens.

_____ **10.** When a river overflows its banks, what is the area along the river that is formed from deposited sediment called?
   a. floodplain
   b. levee
   c. dam
   d. placer deposit

Match the correct description with the correct term. Write the letter in the space provided.

_____ **11.** a sudden flood that can lead to property damage and death

_____ **12.** a barrier that can redirect and hold floodwater in a reservoir

_____ **13.** sediment buildup along a river channel that helps keep a river inside its banks

_____ **14.** used to build artificial levees to control water during flooding

a. sandbags

b. flash flood

c. levee

d. dam

Directed Reading A

Section: Using Water Wisely (pp. 378–383)

Match the correct description with the correct term. Write the letter in the space provided.

_____ 1. percentage of your body that is water

_____ 2. percentage of drinkable water on Earth

_____ 3. percentage of Earth's drinkable water frozen in polar ice caps

a. 3%

b. 65%

c. 75%

GROUNDWATER

Match the correct description with the correct term. Write the letter in the space provided.

_____ 4. water found inside rocks below Earth's surface

_____ 5. the layer of rock or sediment that stores and allows the flow of groundwater

_____ 6. the upper surface of underground water

a. aquifer

b. groundwater

c. water table

WATER IN CALIFORNIA

Write the letter of the correct answer in the space provided.

_____ 7. What is one problem with many California aquifers?
   a. The water is too hot.
   b. The water is frozen.
   c. The water is overflowing.
   d. The water is contaminated.

Match the correct description with the correct term. Write the letter in the space provided.

_____ 8. taking out more groundwater than rain is replacing

_____ 9. a place for storing surface water, such as rain

a. reservoir

b. overdraft

| Directed Reading A *continued*

Where Does It Come From?

Write the letter of the correct answer in the space provided.

_____ **10.** About how much of California's precipitation happens north of
Sacramento?

a. 10%	**c.** 50%
b. 25%	**d.** 75%

_____ **11.** About how much water in California is used south of Sacramento?

a. 10%	**c.** 50%
b. 25%	**d.** 75%

Where Does It Go?

_____ **12.** How much of California's water may be used by agriculture each year?

a. up to 25%	**c.** up to 75%
b. up to 50%	**d.** up to 100%

_____ **13.** The use of water by businesses and households is called

a. agricultural use.

b. environmental management use.

c. urban use.

d. recreational use.

WATER POLLUTION

_____ **14.** Which of the following describes water pollution?

a. the harmful waste in bodies of water

b. harmful chemicals in the air that mix with rain

c. the loss of water due to runoff

d. the water used by businesses and households

Sources of Water Pollution

_____ **15.** What do people need to know to prevent groundwater pollution?

a. how old pollutants are

b. where pollutants come from

c. what lives in the water

d. how concentrated the pollution is

The Clean Water Act of 1972

_____ **16.** What law was passed to make surface water clean enough for fishing
and swimming?

a. the Clean Water Act of 1972

b. the Oil Pollution Act of 1990

c. the Oil Tanker Act of 2015

d. the Environmental Protection Act of 1972

Directed Reading A *continued*

Other Water-Quality Laws

_____ **17.** What laws has the Marine Protection, Research, and Sanctuaries Act of 1972 strengthened?
    a. laws against ocean dumping
    b. laws to improve water quality
    c. laws against polluting rivers
    d. laws to prevent more oil spills

WATER CONSERVATION

_____ **18.** What is the preservation and wise use of natural resources called?
    a. pollution
    b. recycling
    c. discharging
    d. conservation

Conserving Water in Agriculture and Industry

_____ **19.** How do farmers keep from losing water by evaporation and runoff?
    a. by using drip irrigation
    b. by installing low-flow toilets
    c. by building aquifers
    d. by discharging used water

_____ **20.** How do many industries conserve cooling water and wastewater?
    a. by discharging used water
    b. by evaporating used water
    c. by removing used water
    d. by recycling used water

Conserving Water at Home

_____ **21.** How can home owners help conserve household water?
    a. by discharging used water
    b. by watering lawns every day
    c. by building aquifers
    d. by planting native plants

What You Can Do

_____ **22.** What is one good way to conserve water at home?
    a. Take shorter showers.
    b. Plant large gardens.
    c. Run the water all day.
    d. Take longer showers.

Skills Worksheet

Vocabulary and Section Summary A

The Active River
VOCABULARY

In your own words, write a definition of the following terms in the space provided.

1. erosion

_____

_____

2. water cycle

_____

_____

3. tributary

_____

_____

4. watershed

_____

_____

5. divide

_____

_____

6. channel

_____

_____

7. load

_____

_____

SECTION SUMMARY

Read the following section summary.

- Rivers shape Earth's landscape through the process of erosion.

- The sun is the major source of energy that drives the water cycle.

- A river system is made up of a network of streams and rivers.

- A watershed is a region that collects runoff water that then becomes part of a river system that drains into a lake or the ocean.

- Gradient affects the speed at which water flows in a stream. The higher the gradient, the faster the water flows. Water that flows rapidly has a lot of energy for eroding soil and rock.

- When a stream's discharge increases, the stream's erosive energy also increases.

- A stream can carry eroded particles as bed load, suspended load, or dissolved load. A stream that has a load of large particles has a high rate of erosion.

- A river can be described as youthful, mature, old, or rejuvenated based on its characteristics.

Skills Worksheet

Vocabulary and Section Summary A

Stream and River Deposits

VOCABULARY

In your own words, write a definition of the following terms in the space provided.

1. deposition

_____

_____

_____

2. delta

_____

_____

_____

3. alluvial fan

_____

_____

_____

4. floodplain

_____

_____

_____

SECTION SUMMARY

Read the following section summary.

- Sediment forms several types of deposits, such as deltas, alluvial fans, and floodplains.
- A delta is a fan-shaped deposit of sediment that forms where a river meets a large body of water.
- Alluvial fans can form when a river deposits sediment on land.
- Flooding brings rich soil to farmland and may cause a stream to change course.
- Flooding can also lead to property damage and death.

Skills Worksheet

Vocabulary and Section Summary A

Using Water Wisely

VOCABULARY

In your own words, write a definition of the following terms in the space provided

1. aquifer

_____

_____

_____

2. water table

_____

_____

_____

3. water pollution

_____

_____

_____

4. conservation

_____

_____

_____

_____

SECTION SUMMARY

Read the following section summary.

- An aquifer is a rock and soil layer that stores groundwater and allows the flow of groundwater.

- California receives its water from surface water, from aquifers, and from other areas.

- Water sources can be polluted by cities, factories, and farms.

- Water can be conserved by using only the water that is needed, by recycling water, and by using drip irrigation systems.

Skills Worksheet

Directed Reading A

Section: Earth's Oceans (pp. 400–405)
Write the letter of the correct answer in the space provided.

_____ **1.** How much of Earth's surface is covered with water?
   a. 25%
   b. 53%
   c. 71%
   d. 82%

_____ **2.** What divides the global ocean into five main oceans?
   a. rivers
   b. mountain ranges
   c. the equator
   d. the continents

DIVISIONS OF THE GLOBAL OCEAN

_____ **3.** Which of the following is the largest ocean?
   a. Atlantic Ocean
   b. Arctic Ocean
   c. Indian Ocean
   d. Pacific Ocean

_____ **4.** Which is about half the size of the Pacific Ocean?
   a. Atlantic Ocean
   b. Arctic Ocean
   c. Indian Ocean
   d. Pacific Ocean

_____ **5.** Which is the third-largest ocean?
   a. Atlantic Ocean
   b. Arctic Ocean
   c. Indian Ocean
   d. Pacific Ocean

_____ **6.** Which ocean is mostly covered by ice?
   a. Atlantic Ocean
   b. Arctic Ocean
   c. Indian Ocean
   d. Pacific Ocean

CHARACTERISTICS OF OCEAN WATER

Ocean Water Is Salty

_____ **7.** Which of the following statements about salt in the ocean is true?
  a. It has been added to the ocean for only 100 years.
  b. It is saltier than the salt we eat.
  c. It is sodium chloride.
  d. It is less salty than the salt we eat.

_____ **8.** For how long have salts been collected in the ocean?
  a. less than 1,000 years
  b. more than 1,000 years, less than 100,000 years
  c. more than 100,000 years, less than 1 million years
  d. more than 1 billion years

Salinity

_____ **9.** What is salinity?
  a. the amount of dissolved salts in a liquid
  b. the amount of sodium in a liquid
  c. the amount of water that has evaporated
  d. the amount of liquids in a solid

Climate Affects Salinity

_____ **10.** Which waters tend to have a higher salinity?
  a. coastal waters in cool, humid environments
  b. river waters
  c. coastal waters in hot, dry climates
  d. coastal waters near river outlets

_____ **11.** Which of the following happens when water evaporates?
  a. Dissolved salts in the water also evaporate.
  b. Table salt is harvested.
  c. Dissolved salts rise to the surface of the water.
  d. Salts and dissolved salts in the water are left behind.

Water Movement Affects Salinity

_____ **12.** Which of the following has a higher salinity?
  a. water in areas with swift currents
  b. water at high tide
  c. slow-moving areas of water
  d. waterfalls

| Directed Reading A *continued*

TEMPERATURE OF OCEAN WATER

_____ **13.** Which of the following statements is generally true?
  a. Ocean water gets warmer as it gets deeper.
  b. Ocean water gets cooler as it gets deeper.
  c. The temperature of ocean water never changes.
  d. Ocean water is coldest near the surface.

Surface Zone

_____ **14.** Convection currents can transfer heat from the surface zone down to what depth?
  a. 1 m to 100 m
  b. 100 m to 300 m
  c. 300 m to 500 m
  d. 500 m to 1,000 m

Thermocline

_____ **15.** In which of the following does water temperature drop more rapidly with increased depth?
  a. surface zone
  b. thermocline
  c. deep zone
  d. convection current

Deep Zone

_____ **16.** What is the usual temperature of deep zone water?
  a. 0°C
  b. 2°C
  c. 10°C
  d. 20°C

Surface Temperature Changes

_____ **17.** Generally, the surface temperature of ocean water is warmer at which latitudes?
  a. lower latitudes
  b. middle latitudes
  c. higher latitudes
  d. middle and higher latitudes

| Directed Reading A *continued*

_____ **18.** The surface temperature of ocean water at higher latitudes will
generally be warmer at what time of year?
   a. spring
   b. summer
   c. fall
   d. winter

Density

_____ **19.** Which of the following is true about water temperature?
   a. It affects the density of water more than salinity.
   b. It affects the density of water less than salinity.
   c. It affects the density of water the same as salinity.
   d. It does not affect the density of water.

Skills Worksheet

Directed Reading A

Section: The Ocean Floor (pp. 406–411)
STUDYING THE OCEAN FLOOR
Seeing by Sonar

Write the letter of the correct answer in the space provided.

_____ **1.** What does *sonar* stand for?
  a. sound and radar
  b. sound navigation and radio
  c. sub-ocean navigation and ranging
  d. sound navigation and ranging

_____ **2.** What can scientists do with sonar technology?
  a. play music for dolphins
  b. calculate the ocean's depth
  c. measure the length of an ocean
  d. calculate the speed of sound

Underwater Vessels

_____ **3.** What is *Deep Flight*?
  a. a high-flying airplane
  b. an underwater vessel
  c. a home for pilots
  d. a sunken ship

_____ **4.** Why are ROVs sometimes better than piloted vessels?
  a. They are more expensive.
  b. They can explore greater depths.
  c. They are easier to build.
  d. They are shaped like fish.

_____ **5.** What part of the ocean does *JASON II* explore?
  a. the ocean surface
  b. the shoreline
  c. the ocean floor
  d. the mid-level currents

THE INTEGRATED OCEAN DRILLING PROGRAM (IODP)

_____ **6.** Why does the IODP perform ocean drilling?
   a. to find oil
   b. to find natural gas
   c. to vent the ocean floor
   d. to learn about the history of Earth

Studying via Satellite

_____ **7.** What does Geosat measure?
   a. changes in the height of the ocean's surface
   b. changes in the ocean's temperature
   c. the direction and speed of ocean currents
   d. the velocity of waves hitting the shoreline

_____ **8.** How is using satellites to make maps of the ocean floor better than using sonar readings from ships?
   a. Satellites are farther away from the ocean.
   b. Satellites cover much more territory.
   c. Ships are always getting lost.
   d. Satellites are less expensive.

OCEAN-FLOOR BASICS

Two Major Regions of Ocean Floor

Match the correct description with the correct term. Write the letter in the space provided.

_____ **9.** edge of the continent covered by ocean water

_____ **10.** extends under the deepest parts of the ocean

a. deep-ocean basin
b. continental margin

Subdivisions of the Ocean Floor

Write the letter of the correct answer in the space provided.

_____ **11.** Which of the following is not a subdivision of the continental margin?
   a. continental shelf
   b. mid-ocean ridges
   c. continental slope
   d. continental rise

_____ **12.** Which of the following is not a subdivision of the deep-ocean basin?
   a. abyssal plain
   b. mid-ocean ridges
   c. continental slope
   d. ocean trenches

TECTONIC PLATES AND OCEAN-FLOOR FEATURES

_____ **13.** How long is the world's longest mountain chain?
   a. 24,000 km
   b. 50,000 km
   c. 64,000 km
   d. 80,000 km

_____ **14.** What causes tectonic plates to move?
   a. the rotation of Earth
   b. the moon's gravity
   c. deep-ocean currents
   d. convection within Earth

Abyssal Plain

_____ **15.** What are the flattest regions on Earth?
   a. seamounts
   b. abyssal plains
   c. mid-ocean ridges
   d. ocean trenches

Seamounts

_____ **16.** Which of the following is a submerged volcanic mountain?
   a. seamount
   b. abyssal plain
   c. mid-ocean ridge
   d. ocean trench

_____ **17.** What is a volcanic island that has submerged and been eroded by waves?
   a. guyot
   b. abyssal plain
   c. rift
   d. ocean trench

Mid-Ocean Ridges

_____ **18.** Which of the following is a long, undersea mountain chain along the ocean's floor?
- **a.** seamount
- **b.** abyssal plain
- **c.** mid-ocean ridge
- **d.** ocean trench

_____ **19.** Which of the following is a crack in the ocean floor?
- **a.** guyot
- **b.** abyssal plain
- **c.** rift
- **d.** ocean trench

Ocean Trenches

_____ **20.** Which of the following is a long, narrow depression in a deep ocean basin?
- **a.** seamount
- **b.** abyssal plain
- **c.** mid-ocean ridge
- **d.** ocean trench

_____ **21.** What is subduction?
- **a.** when one tectonic plate moves over another
- **b.** when one tectonic plate collides with another
- **c.** when one tectonic plate moves under another
- **d.** when one tectonic plate moves away from another

Directed Reading A

Section: Resources from the Ocean (pp. 412–415)
LIVING RESOURCES
Write the letter of the correct answer in the space provided.

_____ 1. Which statement about ocean resources is true?
  a. Humans get very few resources from the ocean.
  b. Demand for the ocean's resources is increasing.
  c. The ocean is so big that we will never use up its resources.
  d. The ocean's resources have already been used up.

Fishing the Ocean

_____ 2. How much fish is harvested from the ocean each year?
  a. almost 7 million tons
  b. almost 17 million tons
  c. almost 75 million tons
  d. almost 700 million tons

_____ 3. What is overfishing?
  a. catching fish with drift nets
  b. catching more fish than can be naturally replaced
  c. catching fish that weigh under 10 pounds
  d. fishing over the neritic zone

Farming the Ocean

_____ 4. Why is overfishing a problem?
  a. It reduces the fish population.
  b. It leaves the fish population as it is.
  c. It increases the fish population.
  d. It results in holding ponds.

_____ 5. Why do people raise fish in fish farms?
  a. Many ponds are available.
  b. Fish farms help meet humans' demand for seafood.
  c. Fish farms are easy to operate.
  d. Fish farms produce better fish than the ocean.

_____ 6. Why would seaweed be a good addition to your diet?
  a. It is high in vitamin C.
  b. It is easy to digest.
  c. It is high in sugar.
  d. It is high in protein.

NONLIVING RESOURCES

Match the correct description with the correct term. Write the letter in the space provided.

_____ **7.** generated by the movement of tides

_____ **8.** removing salt from sea water

_____ **9.** formed from the remains of living things

_____ **10.** renewed by the water cycle in most countries

a. desalination

b. fresh water

c. tidal energy

d. oil

Match the correct description with the correct term. Write the letter in the space provided.

_____ **11.** crystallized lumps of minerals

_____ **12.** found between layers of impermeable rock

a. natural gas

b. nodules

Skills Worksheet

Directed Reading A

Section: Ocean Pollution (pp. 416–421)

Write the letter of the correct answer in the space provided.

_____ 1. Which of the following statements is true?

    a. Humans do not throw trash into the oceans.

    b. Humans have thrown trash into the oceans for hundreds or thousands of years.

    c. Humans started throwing trash in the oceans only a few years ago.

    d. Trash in the oceans is not harmful to plants, animals, or people.

NONPOINT-SOURCE POLLUTION

_____ 2. Why is nonpoint-source pollution difficult to trace?

    a. It is easily identified.

    b. It comes from one specific source.

    c. It is less harmful than other kinds of pollution.

    d. It comes from many sources rather than one.

_____ 3. How can people reduce nonpoint-source pollution?

    a. use boats only in the ocean

    b. dispose of used motor oil properly

    c. use more lawn fertilizer

    d. use more pesticides

POINT-SOURCE POLLUTION

_____ 4. Which of the following statements is true about point-source pollution?

    a. It comes from unidentified sources.

    b. It comes from a specific source.

    c. It is harmful only to plants.

    d. It is not a problem today.

Trash Dumping

_____ 5. What harmful trash was washing up on beaches in the 1980s?

    a. vials of blood and needles

    b. seaweed

    c. mussel shells

    d. old newspapers

| Directed Reading A *continued*

Effects of Trash Dumping

_____. **6.** Why is plastic trash harmful to marine animals?

    a. Plastic trash breaks down too quickly.

    b. Animals get tangled in plastic trash and can be strangled.

    c. Plastic trash is salty.

    d. Plastic trash floats and blocks sunlight.

Sludge Dumping

_____ **7.** What is sludge?

    a. bacteria

    b. a sewer drain

    c. the solid part of waste

    d. liquid and solid waste

Oil Spills

_____ **8.** What can happen if oil shipments are not handled properly?

    a. The price of oil will go down.

    b. The oil can be easily cleaned up.

    c. Airplanes will have to transport oil.

    d. The oil can spill into the ocean.

Effects of Oil Spills

_____ **9.** Which of the following is an effect of an oil spill?

    a. Many plants and animals die.

    b. The fishing industry improves.

    c. Plant life increases.

    d. The oil cleans the beaches.

Preventing Oil Spills

_____ **10.** How are oil tankers being built to prevent oil spills?

    a. They have rubber hulls.

    b. They have two hulls.

    c. They go much faster.

    d. They have reef bumpers.

Directed Reading A *continued*

SAVING OUR OCEAN RESOURCES

Match the correct description with the correct term. Write the letter in the space provided.

_____ **11.** law that regulates ocean dumping permits

_____ **12.** law that prohibits dumping harmful materials into the ocean

_____ **13.** one of the largest beach clean-up programs in the United States

a. Adopt-a-Beach

b. Clean Water Act

c. U.S. Marine Protection, Research, and Sanctuaries Act

Match the correct description with the correct term. Write the letter in the space provided.

_____ **14.** People around the world have demanded that governments do more about this.

_____ **15.** A treaty signed by sixty-four countries in 1989 banned this.

a. ocean dumping

b. ocean pollution

Skills Worksheet

Vocabulary and Section Summary A

Earth's Oceans

VOCABULARY

In your own words, write a definition of the following terms in the space provided.

1. salinity

_____

_____

_____

2. thermocline

_____

_____

_____

SECTION SUMMARY

Read the following section summary.

- The global ocean is divided by the continents into five main oceans: Pacific Ocean, Atlantic Ocean, Indian Ocean, Southern Ocean, and Arctic Ocean.

- Salts have collected in the ocean for billions of years. Salinity is a measure of the amount of dissolved salts in a given mass of liquid.

- The three temperature zones of ocean water are the surface zone, the thermocline, and the deep zone.

- Temperature and salinity determine the density of ocean water. The density of ocean water drives convection currents.

Vocabulary and Section Summary A

The Ocean Floor
VOCABULARY

In your own words, write a definition of the following terms in the space provided.

1. abyssal plain

_____

_____

_____

2. mid-ocean ridge

_____

_____

_____

3. ocean trench

_____

_____

_____

SECTION SUMMARY

Read the following section summary.

- Scientists study the ocean floor by using sonar, underwater vessels, drilling, and satellites.
- The ocean floor is divided into two regions—the continental margin and the deep-ocean basin.
- The continental margin consists of the continental shelf, the continental slope, and the continental rise.
- The deep-ocean basin consists of the abyssal plain, mid-ocean ridges, trenches, and seamounts.
- Mid-ocean ridges and trenches mark the boundaries of tectonic plates.

Vocabulary and Section Summary A

Resources from the Ocean

VOCABULARY

In your own words, write a definition of the following term in the space provided.

1. desalination

_____

_____

_____

SECTION SUMMARY

Read the following section summary.

- Humans depend on the ocean for living and nonliving resources.

- Fish and other marine life are caught in the ocean and are being raised in fish farms to help feed growing human populations.

- Nonliving ocean resources include oil and natural gas, fresh water, minerals, and tidal energy.

Vocabulary and Section Summary A

Ocean Pollution
VOCABULARY
In your own words, write a definition of the following terms in the space provided.

1. nonpoint-source pollution

_____

_____

_____

2. point-source pollution

_____

_____

_____

SECTION SUMMARY
Read the following section summary.

- The two main types of water pollution are nonpoint-source pollution and pointsource pollution.

- Types of nonpoint-source pollution include oil and gasoline from cars, trucks, and watercraft, as well as of pesticides, herbicides, and fertilizers.

- Oil spills harm wildlife and local fishing economies and cost billions of dollars to clean up.

- Efforts to save ocean resources include laws, international treaties, and volunteer cleanups.

Skills Worksheet

Directed Reading A

Section: Currents (pp. 436–443)

Write the letter of the correct answer in the space provided.

_____ **1.** What are streamlike movements of water in the ocean called?
  a. ocean crests
  b. continental deflections
  c. the Coriolis effect
  d. ocean currents

SURFACE CURRENTS

_____ **2.** What are movements of water near the ocean's surface called?
  a. surface currents
  b. deep currents
  c. global currents
  d. convection currents

_____ **3.** Which of the following does NOT control surface currents?
  a. ocean traffic
  b. continental deflections
  c. global winds
  d. the Coriolis effect

Global Winds

_____ **4.** In what direction do winds near the equator blow ocean water?
  a. west to east
  b. east to west
  c. toward the land
  d. toward the open sea

How the Sun Powers Ocean Currents

_____ **5.** Which of the following is the major source of energy that causes surface currents to form?
  a. the stars
  b. the equator
  c. the moon
  d. the sun

| Directed Reading A *continued* |

The Coriolis Effect

_____ **6.** Because of the Coriolis effect, what happens to ocean currents?
   a. They move in curved paths.
   b. They move in straight paths.
   c. They move toward the land.
   d. They move toward the open sea.

_____ **7.** Where are currents deflected in the Northern Hemisphere?
   a. to the left
   b. to the north
   c. to the right
   d. to the south

Continental Deflections

_____ **8.** What happens to an ocean current when it meets a continent?
   a. It gets hot.
   b. It sinks.
   c. It moves faster.
   d. It changes direction.

How Surface Currents Distribute Heat

_____ **9.** How do surface currents transfer heat energy?
   a. by convection
   b. by the Coriolis effect
   c. by a trough
   d. by upwelling

DEEP CURRENTS

Match the correct description with the correct term. Write the letter in the space provided.

_____ **10.** a movement of ocean water located far below the surface

_____ **11.** the amount of matter in a given space

_____ **12.** the amount of dissolved salts or solids in a liquid

a. deep current

b. density

c. salinity

How Deep Currents Form

Write the letter of the correct answer in the space provided.

_____ **13.** What happens to dense ocean water?
  a. It rises.
  b. It gets hot.
  c. It sinks.
  d. It gets cold.

CONVECTION CURRENTS

_____ **14.** What are surface currents and deep currents together called?
  a. global winds
  b. convection currents
  c. Coriolis effect
  d. continental deflections

GLOBAL CIRCULATION

Match the correct description with the correct term. Write the letter in the space provided.

_____ **15.** brought to the surface as deep water rises

_____ **16.** taken from the surface to the deep ocean

_____ **17.** carried by convection currents

a. oxygen

b. heat

c. nutrients

Directed Reading A

Section: Currents and Climate (pp. 444–447)
SURFACE CURRENTS AND CLIMATE
Write the letter of the correct answer in the space provided.

_____ 1. What does the surface temperature of the water affect?
    a. inland areas
    b. the air above it
    c. the deep water
    d. convection currents

Warm-Water Currents and Climate

_____ 2. Where does the Gulf Stream gets its heat to warm Great Britain?
    a. from the Tropics
    b. from the North Atlantic
    c. from the Scilly Isles
    d. from the poles

Cold-Water Currents and Climate

_____ 3. In what direction does the California Current carry cold water?
    a. north
    b. south
    c. east
    d. west

EL NIÑO AND LA NIÑA
Match the correct description with the correct term. Write the letter in the space provided.

_____ 4. a change in the water temperature in the Pacific Ocean that makes a warm current

    a. La Niña
    b. El Niño

_____ 5. a change in the eastern Pacific Ocean in which the surface water temperature becomes unusually cool

Directed Reading A *continued*

El Niño and Weather Patterns

Write the letter of the correct answer in the space provided.

_____ **6.** What should scientists study to predict the weather changes that are caused by El Niño?
   a. the sun and atmosphere
   b. the ocean and stars
   c. the stars and moon
   d. the atmosphere and ocean

Studying and Predicting El Niño

_____ **7.** Which of the following can scientists use to collect data to predict El Niño?
   a. buoys
   b. tides
   c. clocks
   d. currents

Effects of El Niño

_____ **8.** What is a very long period of time in which there isn't enough rain and which can lead to crop failure?
   a. mudslide
   b. life span
   c. drought
   d. tsunami

UPWELLING

_____ **9.** What is it called when deep, cold, nutrient-rich water comes to the surface?
   a. convection current
   b. La Niña
   c. Coriolis effect
   d. upwelling

Skills Worksheet

Directed Reading A

Section: Waves and Tides (pp. 448–455)

WAVES

Write the letter of the correct answer in the space provided.

_____ 1. Which of the following is a disturbance that carries energy through matter?
  a. wave
  b. wind
  c. heat
  d. climate

_____ 2. What determines the size of an ocean wave?
  a. the wave's color
  b. the wave's temperature
  c. the wave's speed
  d. the wave's energy

Parts of a Wave

Match the correct description with the correct term. Write the letter in the space provided.

_____ 3. the highest point of a wave   **a.** crest

_____ 4. the lowest point of a wave   **b.** trough

Match the correct description with the correct term. Write the letter in the space provided.

_____ 5. the distance between two crests or troughs   **a.** wave height

_____ 6. the vertical distance between a crest and a trough   **b.** wavelength

Wave Formation

Write the letter of the correct answer in the space provided.

_____ **7.** What is the source of energy that forms most ocean waves?
  a. wind
  b. earthquakes
  c. stars
  d. water

Wave Movement

_____ **8.** How does water move when a wave of energy passes through it?
  a. forward
  b. backward
  c. into the air
  d. in circular movements

Wave Energy

_____ **9.** What happens when the wind keeps blowing across the water in the same direction?
  a. The waves move slower.
  b. The waves stay the same size.
  c. The waves get larger.
  d. The waves get smaller.

Wave Speed

_____ **10.** What do you need to know to determine wave speed?
  a. wavelength and wave height
  b. wavelength and wave period
  c. wave energy and wave height
  d. wavelength and wave temperature

_____ **11.** What does a wave period measure?
  a. speed of each wave
  b. wave energy
  c. time between two waves
  d. wave speed

_____ **12.** When waves reach the shore, what is transferred to the beach?
  a. El Niño
  b. energy
  c. cold
  d. oxygen

Why Waves Break

_____ **13.** Which of the following is a high wave that crashes down onto the ocean floor?
a. breaker
b. tide
c. upwelling
d. deflection

TIDES

_____ **14.** What are daily changes in the level of ocean water called?
a. tides
b. floods
c. waves
d. currents

Why Tides Happen

_____ **15.** Why do we notice the moon's gravitational pull more on liquids than on solids?
a. Solids weigh more.
b. Solids move more easily.
c. Liquids weigh less.
d. Liquids move more easily.

Where Tides Happen

_____ **16.** Which of the following is pulled toward the moon with the greatest force?
a. the part of the ocean on Earth's far side
b. the poles
c. the part of the ocean that directly faces the moon
d. the solid center of Earth

High Tides and Low Tides

_____ **17.** What are the bulges that form because of the moon's gravity called?
a. low tides
b. high tides
c. wave speed
d. tsunamis

_____ **18.** What forms between the high tides?
   a. tsunamis
   b. wave speed
   c. neap tides
   d. low tides

Timing the Tides

_____ **19.** How long does it take a spot on Earth that is facing the moon to rotate until it faces the moon again?
   a. 50 h and 24 m
   b. 24 h and 50 m
   c. 24 h
   d. 50 m

TIDAL VARIATIONS

_____ **20.** What is the difference between levels of ocean water at high tide and low tide called?
   a. tidal wave
   b. neap tide
   c. spring tide
   d. tidal range

Spring Tides

_____ **21.** What are tides that happen during the new and full moons called?
   a. high tides
   b. neap tides
   c. spring tides
   d. low tides

Neap Tides

_____ **22.** What are tides that happen during the first and third quarters of the moon called?
   a. high tides
   b. neap tides
   c. spring tides
   d. low tides

Skills Worksheet

Vocabulary and Section Summary A

Currents

VOCABULARY

In your own words, write a definition of the following terms in the space provided.

1. surface current

_____

_____

2. Coriolis effect

_____

_____

3. deep current

_____

_____

4. convection current

_____

_____

SECTION SUMMARY

Read the following section summary.

• Surface currents form as wind transfers energy to ocean water.

• Surface currents are controlled by three factors: global winds, the Coriolis effect, and continental deflections.

• Deep currents form where the density of ocean water increases. Water density depends on temperature and salinity.

• Surface currents and deep currents combine to form convection currents that transfer energy.

• Earth's global circulation moves water through all oceans and distributes materials and heat.

Vocabulary and Section Summary A

Currents and Climate

VOCABULARY

In your own words, write a definition of the following terms in the space provided.

1. El Niño

_____

_____

2. La Niña

_____

_____

3. upwelling

_____

_____

SECTION SUMMARY

Read the following section summary.

• Surface currents cause climates of coastal areas to be more moderate than inland climates at the same latitude and elevation.

• Upwelling is the flow of cold, nutrient-rich water from the deep ocean to the surface.

• During El Niño, warm and cool surface waters change locations. El Niño can cause floods, mudslides, drought, and changes in upwelling.

Skills Worksheet

Vocabulary and Section Summary A

Waves and Tides

VOCABULARY

In your own words, write a definition of the following terms in the space provided.

1. wave

_____

_____

2. tide

_____

_____

3. spring tide

_____

_____

4. neap tide

_____

_____

SECTION SUMMARY

Read the following section summary.

- Waves form when the wind's energy is transferred to the surface of the ocean.

- Wave energy travels through water near the water's surface, while the water itself rises and falls in circular movements.

- Waves break when the water depth becomes so shallow that the bottom of the wave transfers energy to the ocean bottom and the shore.

- Tides are caused by the gravitational forces of the moon and the sun on Earth. The moon's gravity is the main force behind the tides.

- The positions of the sun and moon relative to the position of Earth cause tidal ranges.

Skills Worksheet

Directed Reading A

Section: Characteristics of the Atmosphere (pp. 470–473)

Write the letter of the correct answer in the space provided.

_____ **1.** What is Earth's atmosphere?
   a. oxygen and sunlight
   b. a mixture of particles
   c. a mixture of gases
   d. water vapor

THE COMPOSITION OF THE ATMOSPHERE

_____ **2.** Earth's atmosphere is made up mostly of which gas?
   a. oxygen
   b. nitrogen
   c. carbon dioxide
   d. argon

_____ **3.** How much of the atmosphere is oxygen?
   a. about 1%
   b. about 78%
   c. about 20%
   d. about 100%

_____ **4.** Where is most of the water in the atmosphere?
   a. in rain
   b. in ice
   c. in water vapor
   d. in carbon dioxide

AIR PRESSURE AND TEMPERATURE

_____ **5.** At sea level, a square inch of Earth is under how much air?
   a. 150 lbs
   b. 15 lbs
   c. 30 lbs
   d. 1500 lbs

Altitude and Air Pressure

_____ **6.** What pulls the gas molecules in the atmosphere toward Earth's surface?
  a. air pressure
  b. gravity
  c. water
  d. solar energy

_____ **7.** What is the measure of the force with which air molecules push on a surface?
  a. Earth's surface
  b. altitude
  c. water vapor
  d. air pressure

_____ **8.** Where is air pressure strongest?
  a. on a mountain top
  b. at Earth's surface
  c. in outer space
  d. close to the sun

Atmospheric Composition and Air Temperature

_____ **9.** Which of the following do warmer layers of the atmosphere have more of than cold layers?
  a. gases that make heat
  b. gases that absorb water
  c. gases that absorb cold
  d. gases that absorb energy from the sun

LAYERS OF THE ATMOSPHERE
The Troposphere: The Layer in Which We Live

_____ **10.** Where does the troposphere lie?
  a. on mountaintops
  b. in the thermosphere
  c. next to Earth's surface
  d. at the ocean bottom

Directed Reading A *continued*

The Stratosphere: Home of the Ozone Layer

_____ **11.** Where is the stratosphere?
  a. within the troposphere
  b. below the troposphere
  c. above the troposphere
  d. below the mesosphere

_____ **12.** What does the ozone layer absorb to protect life?
  a. gas mixtures
  b. ultraviolet rays
  c. particle mixtures
  d. water vapor

The Mesosphere: The Middle Layer

_____ **13.** As altitude increases in the mesosphere, what does temperature do?
  a. It increases.
  b. It decreases.
  c. It remains unchanged.
  d. It cannot be measured.

The Thermosphere: The Edge of the Atmosphere

_____ **14.** What do nitrogen and oxygen atoms in the thermosphere do to solar radiation?
  a. absorb it
  b. emit it
  c. reflect it
  d. destroy it

Skills Worksheet

Directed Reading A

Section: Atmospheric Heating (pp. 474–479)
RADIATION: ENERGY TRANSFER BY WAVES
Write the letter of the correct answer in the space provided.

_____ **1.** How long does it take the sun's energy to reach Earth?
- **a.** about 8 hours
- **b.** about 80 hours
- **c.** about 8 minutes
- **d.** about 8 days

_____ **2.** What is the transfer of energy as electromagnetic waves, including those hitting Earth, called?
- **a.** conduction
- **b.** radiation
- **c.** reflection
- **d.** absorption

The Electromagnetic Spectrum

_____ **3.** What is the name for all frequencies or wavelengths of electromagnetic waves?
- **a.** solar energy
- **b.** ultraviolet radiation
- **c.** electromagnetic spectrum
- **d.** reflected sunlight

_____ **4.** What makes the kinds of electromagnetic radiation differ?
- **a.** wavelength
- **b.** radiation
- **c.** absorption
- **d.** reflection

The Atmosphere and Solar Radiation

_____ **5.** What results when solar radiation reaches Earth's atmosphere?
- **a.** Some becomes absorbed.
- **b.** All becomes absorbed.
- **c.** None becomes absorbed.
- **d.** All becomes ozone.

_____ **6.** What is the form of most solar energy that reaches Earth's surface?
- **a.** ultraviolet waves
- **b.** visible light
- **c.** radio waves
- **d.** infrared radiation

Directed Reading A *continued*

CONDUCTION: ENERGY TRANSFER BY CONTACT

_____ **7.** How is heat transferred in conduction?
    a. by heat waves   **c.** by wavelengths
    b. by heat energy   **d.** by physical contact

_____ **8.** In conduction, where is the kinetic energy of atoms in a hot object transferred?
    a. to hot objects
    b. to cold objects
    c. to potential energy
    d. to solar energy

CONVECTION: ENERGY TRANSFER BY MOTION

_____ **9.** How does convection cause heat transfer to take place in a liquid or gas?
    a. by radiation   **c.** by conduction
    b. by evaporation   **d.** by circulation

_____ **10.** What is the cycle of warm fluid rising and cool fluid sinking called?
    a. radiation current
    b. convection current
    c. conduction current
    d. density current

THE GREENHOUSE EFFECT

_____ **11.** What is the process where atmospheric gases absorb and reradiate solar energy called?
    a. thermal recycling   **c.** global warming
    b. thermal balance   **d.** greenhouse effect

The Radiation Balance: Energy In, Energy Out

_____ **12.** How does the radiation balance affect Earth?
    a. It causes global warming.
    b. It prevents air pollution.
    c. It makes Earth livable.
    d. It circulates solar energy.

GLOBAL WARMING

_____ **13.** What is one possible cause of increased average global temperatures?
    a. radiating solar energy
    b. circulating thermal energy
    c. burning fossil fuels
    d. reducing greenhouse gases

Skills Worksheet

Directed Reading A

Section: Air Movement and Wind (pp. 480–483)
WHAT CAUSES WIND?
Write the letter of the correct answer in the space provided.

_____ 1. What causes differences in air pressure?
  a. unequal heating of Earth
  b. unequal oxygen on Earth
  c. equal heating of Earth
  d. equal temperature of Earth

_____ 2. What causes wind?
  a. differences in water
  b. differences in gases
  c. differences in air pressure
  d. differences in oxygen

_____ 3. What are the large, circular patterns air travels in called?
  a. pressure belts
  b. pressure cells
  c. convection belts
  d. convection cells

The Coriolis Effect

_____ 4. What is the Coriolis effect, the curving of winds and ocean currents, due to?
  a. Earth's rotation
  b. Earth's surface area
  c. Earth's gravity
  d. Earth's atmosphere

_____ 5. In what hemisphere does the Coriolis effect cause northbound winds to travel east?
  a. Southern
  b. Northern
  c. Eastern
  d. Western

GLOBAL WINDS

_____ **6.** With what do convection cells, pressure belts, and winds combine to
form global winds?
   a. Earth's gravity
   b. Coriolis effect
   c. thermal patterns
   d. pressure patterns

LOCAL WINDS

Use the terms from the following list to complete the sentences below.

   sea breeze   valley breeze   local wind
   land breeze   mountain breeze

7. Air flow caused by the properties of Earth's surface matter is

   a(n) _____.

8. Cool air that flows over the ocean toward land during the day is called

   a(n) _____.

9. Cool air that flows over the land toward the ocean during the night is

   called a(n) _____.

10. Warm air that rises up mountain slopes during the day is called

   a(n) _____.

11. Cool air that moves down mountain slopes during the night is

   called a(n) _____.

Skills Worksheet

Directed Reading A

Section: The Air We Breathe (pp. 484–491)
AIR POLLUTION
Write the letter of the correct answer in the space provided.

_____ **1.** What is the introduction of pollutants from human and natural sources into the atmosphere called?
  a. water pollution
  b. air pollution
  c. ground pollution
  d. thermal pollution

Primary Pollutants

_____ **2.** What are pollutants that are put directly into the air by human or natural activity?
  a. secondary pollutants
  b. primary pollutants
  c. natural pollutants
  d. animal pollutants

Secondary Pollutants

_____ **3.** What results when primary pollutants react with other primary pollutants or with naturally occurring substances?
  a. inferior pollutants
  b. reactant pollutants
  c. secondary pollutants
  d. original pollutants

_____ **4.** What is an example of a secondary pollutant?
  a. oxygen
  b. vapor
  c. pollen
  d. ozone

The Formation of Smog

_____ **5.** What do ozone and vehicle exhaust react with to form smog?
  a. sunlight
  b. ultraviolet energy
  c. dust particles
  d. smoke

Directed Reading A *continued*

HUMAN-CAUSED AIR POLLUTION

_____ **6.** How much human-caused air pollution in the United States comes from cars?
  a. about 10% to 20%
  b. about 25% to 35%
  c. about 40% to 50%
  d. about 60% to 70%

Industrial Air Pollution

_____ **7.** What is a very big cause of industrial air pollution?
  a. ozone plants
  b. forest fires
  c. car exhaust
  d. burning fossil fuels

Indoor Air Pollution

_____ **8.** What is one way to reduce indoor air pollution?
  a. Use less fossil fuels.
  b. Use more fossil fuels.
  c. Use chemical solvents.
  d. Improve ventilation.

ACID PRECIPITATION

_____ **9.** What do high concentrations of acids in rain, sleet, or snow help to cause?
  a. ozone rain
  b. acid precipitation
  c. dirty snow
  d. nitric imbalance

Acid Precipitation and Plants

_____ **10.** Which of the following is an effect of acidification?
  a. decreased soil acidity
  b. increased plant nutrients
  c. unchanged soil chemistry
  d. increased soil acidity

The Effects of Acid Precipitation on Forests

_____ 11. What has acid precipitation done to some forests?
  a. It has made them stronger.
  b. It has damaged large areas.
  c. It hasn't done anything.
  d. It has made them grow.

Acid Precipitation and Aquatic Ecosystems

_____ 12. What is a very quick change in acidity in a body of water called?
  a. basic shock
  b. toxic shock
  c. acid shock
  d. ozone shock

THE OZONE HOLE

_____ 13. What does the ozone hole enable to reach Earth's surface?
  a. acid rain
  b. primary pollutants
  c. convection currents
  d. ultraviolet radiation

Cooperation to Reduce the Ozone Hole

_____ 14. What kind of molecules made the ozone layer break down?
  a. oxygen
  b. BHB
  c. CFC
  d. nitrogen

AIR POLLUTION AND HUMAN HEALTH

_____ 15. What are some of the bad effects air pollution has on people?
  a. lung cancer, coughing
  b. sleepiness, baldness
  c. deafness, weight gain
  d. foot problems, weight loss

CLEANING UP AIR POLLUTION

_____ 16. What did the Clean Air Act give the EPA the power to do?
  a. control water pollution levels
  b. control air pollution levels
  c. control ultraviolet radiation
  d. control oxygen levels

Controlling Air Pollution from Industry

_____ **17.** What device is used to take out some pollutants from power plants?
   a. sponges
   b. scrapers
   c. vacuums
   d. scrubbers

The Allowance Trading System

_____ **18.** What does the Allowance Trading System help companies to do?
   a. reduce pollution
   b. trade stock
   c. increase investments
   d. increase acidification

Reducing Air Pollution from Vehicles

_____ **19.** What two kinds of power do hybrid cars use?
   a. water and gasoline
   b. gasoline and electricity
   c. electricity and oxygen
   d. steam and gasoline

Skills Worksheet

Vocabulary and Section Summary A

Characteristics of the Atmosphere

VOCABULARY

In your own words, write a definition of the following terms in the space provided.

1. atmosphere

_____

_____

2. air pressure

_____

_____

3. troposphere

_____

_____

4. stratosphere

_____

_____

5. mesosphere

_____

_____

6. thermosphere

_____

_____

Vocabulary and Section Summary A *continued*

SECTION SUMMARY

Read the following section summary.

- Nitrogen and oxygen make up most of Earth's atmosphere.
- Air pressure decreases as altitude increases.
- The composition of atmospheric layers affects their temperature.
- The troposphere is the lowest atmospheric layer. It is the layer in which we live.
- The stratosphere contains the ozone layer, which protects us from harmful ultraviolet radiation.
- The mesosphere is the coldest atmospheric layer.
- The thermosphere is the uppermost layer of the atmosphere.

Skills Worksheet

Vocabulary and Section Summary A

Atmospheric Heating
VOCABULARY
In your own words, write a definition of the following terms in the space provided.

1. radiation

_____

_____

2. electromagnetic spectrum

_____

_____

3. conduction

_____

_____

4. convection

_____

_____

5. convection current

_____

_____

6. greenhouse effect

_____

_____

SECTION SUMMARY

Read the following section summary.

- Energy travels from the sun to Earth by radiation. This energy drives many processes at Earth's surface.

- Energy in Earth's atmosphere is transferred by radiation, conduction, and convection.

- Radiation is the transfer of energy through space or matter by waves.

- Conduction is the transfer of energy by direct contact.

- Convection is energy transfer by the movement of matter.

Skills Worksheet

Vocabulary and Section Summary A

Air Movement and Wind
VOCABULARY
In your own words, write a definition of the following terms in the space provided.

1. wind

_____

_____

_____

2. Coriolis effect

_____

_____

_____

SECTION SUMMARY
Read the following section summary.

• Winds blow from areas of high pressure to areas of low pressure.

• Pressure belts are caused by the uneven heating of Earth's surface by the sun.

• The Coriolis effect causes wind to appear to curve as it moves across Earth's surface.

• Global winds include the polar easterlies, the westerlies, and the trade winds.

• Local winds include sea and land breezes and valley and mountain breezes.

Vocabulary and Section Summary A

The Air We Breathe
VOCABULARY
In your own words, write a definition of the following terms in the space provided.

1. air pollution

_____

_____

_____

2. acid precipitation

_____

_____

_____

SECTION SUMMARY
Read the following section summary.

• Air pollution is the introduction of harmful substances into the air by humans or by natural events.

• Primary pollutants are pollutants that are put directly into the air by human or natural activity.

• Secondary pollutants are pollutants that form when primary pollutants react with other primary pollutants or with naturally occurring substances.

• Transportation, industry, and natural sources are the main sources of air pollution.

• The burning of fossil fuels may lead to air pollution and acid precipitation, which may harm human and wildlife habitats.

• Air pollution can be reduced by legislation, such as the Clean Air Act; by technology, such as scrubbers; and by changes in lifestyle.

Skills Worksheet

Directed Reading A

Section: Water in the Air (pp. 506–511)

Write the letter of the correct answer in the space provided.

_____ **1.** What is the condition of the atmosphere at a certain time and place called?
  a. water cycle
  b. humidity
  c. weather
  d. dew point

THE WATER CYCLE

_____ **2.** What heats Earth's surface and causes water to change states?
  a. moon
  b. stars
  c. equator
  d. sun

_____ **3.** What is the movement of water between the atmosphere, the land, and the oceans called?
  a. condensation
  b. weather
  c. water cycle
  d. evaporation

HUMIDITY

Use the terms from the following list to complete the sentences below.

  humidity   vapor pressure   dew point

4. The temperature at which air is saturated is the _____.

5. The rate of condensation increases as _____ increases.

6. The amount of water vapor in the air is called _____.

Relative Humidity

Write the letter of the correct answer in the space provided.

_____ **7.** What is a measure of how close the air is to the dew point called?
  a. dew point
  b. relative humidity
  c. vapor pressure
  d. weather

Measuring Relative Humidity

_____ **8.** What does a psychrometer do?
   a. It measures humidity.
   b. It measures temperature.
   c. It measures condensation.
   d. It measures precipitation.

CONDENSATION

_____ **9.** What is it called when a gas changes to a liquid?
   a. evaporation
   b. condensation
   c. humidity
   d. water vapor

_____ **10.** When do liquid water droplets form?
   a. when the air is saturated
   b. when the relative humidity goes down
   c. when the air is very warm
   d. when the relative humidity is 50%

Dew Point and Condensation

_____ **11.** What does NOT happen when ice cubes are added to a glass of water?
   a. The temperature of the water drops.
   b. The glass absorbs heat from the air.
   c. The temperature of the water rises.
   d. The cold water absorbs heat from the glass.

Reaching the Dew Point

_____ **12.** What are water droplets that form on grass during the night called?
   a. vapor
   b. dew
   c. hail
   d. rain

CLOUDS AND PRECIPITATION

_____ **13.** What is a cloud made of?
   a. gases
   b. warm air
   c. water droplets
   d. snow

Directed Reading A *continued*

_____ **14.** What forms in a cloud when temperatures are below 0°C?
   a. gases
   b. water vapor
   c. water droplets
   d. ice crystals

Types of Clouds

_____ **15.** How are clouds classified?
   a. by their shape and altitude
   b. by their altitude and temperature
   c. by their shape and color
   d. by their temperature and shape

_____ **16.** Which of the following is NOT a cloud form?
   a. stratus
   b. air mass
   c. cumulus
   d. cirrus

Precipitation

_____ **17.** Which best defines precipitation?
   a. water changing into water vapor
   b. water falling to Earth
   c. water flowing across land
   d. water going from Earth into the air

_____ **18.** What is the most common form of precipitation?
   a. rain
   b. snow
   c. sleet
   d. hail

Directed Reading A

Section: Fronts and Weather (pp. 512–519)

Write the letter of the correct answer in the space provided.

_____ **1.** What causes the weather to change?
   a. Earth warms up.
   b. Air masses interact.
   c. Clouds appear.
   d. The air vibrates.

_____ **2.** What is a large body of air with the same temperature and moisture content throughout called?
   a. air mass
   b. front
   c. cyclone
   d. cloud

FRONTS

_____ **3.** What happens when a warm air mass meets a cold air mass?
   a. They usually mix together.
   b. The warm air rises.
   c. The cold air rises.
   d. They always remain apart.

_____ **4.** What is the area where different air masses meet called?
   a. front
   b. tornado
   c. severe weather
   d. condensation

Cold Fronts

Match the correct description with the correct term. Write the letter in the space provided.

_____ **5.** Warm air moves over cold, denser air.

_____ **6.** A cold air mass meets a warm air mass, but both air masses remain separated.

_____ **7.** A warm air mass is caught between two colder air masses.

_____ **8.** Cold air moves under warm air, pushing the warm air up.

   a. cold front
   b. warm front
   c. occluded front
   d. stationary front

Directed Reading A *continued*

Match the correct description with the correct term. Write the letter in the space provided.

_____ **9.** drizzly rain; then clear, warm weather

_____ **10.** many days of cloudy, wet weather

_____ **11.** heavy rain or snow

_____ **12.** cool temperatures; large amounts of rain and snow

a. cold front

b. warm front

c. occluded front

d. stationary front

AIR PRESSURE AND WEATHER

Write the letter of the correct answer in the space provided.

_____ **13.** What kind of air pressure does a cyclone have?
  a. lower than surrounding areas
  b. higher than surrounding areas
  c. sinking air pressure
  d. increasing air pressure

_____ **14.** What happens as the air in a cyclone's center rises?
  a. The air loses energy.
  b. The air cools.
  c. The air sinks.
  d. The air gets warmer.

_____ **15.** What are areas of high pressure called?
  a. dew point
  b. cyclones
  c. lightning
  d. anticyclones

_____ **16.** What kind of weather does an anticyclone bring?
  a. humid
  b. snowy
  c. dry and clear
  d. cloudy and wet

THUNDERSTORMS

_____ **17.** Which of the following best describes a thunderstorm?
  a. heavy snow
  b. cold and dry
  c. strong winds, heavy rain, and lightning
  d. decreasing humidity

_____ **18.** Which of the following is a condition that produces a thunderstorm?
   a. cold air near Earth and an unstable atmosphere
   b. warm, moist air near Earth's surface and an unstable atmosphere
   c. cold air high in the atmosphere
   d. dry air and a stable atmosphere

_____ **19.** Which kind of cloud makes a thunderstorm?
   a. stratus
   b. cirrus
   c. cumulus
   d. cumulonimbus

Lightning

Use the terms from the following list to complete the sentences below.

   thunder   lightning

20. The electric discharge between two clouds is called _____.

21. The sound caused by air rapidly expanding along an electrical strike is

   called _____.

TORNADOES

Write the letter of the correct answer in the space provided.

_____ **22.** What is a tornado?
   a. a rapidly, spinning column of air that touches the ground
   b. a rapid expansion of air
   c. a brief, light shower
   d. a sudden, severe thunderstorm

_____ **23.** How does a tornado begin?
   a. The air is cool and dry.
   b. A funnel cloud forms.
   c. Winds make a layer of air spin.
   d. Lightning releases energy.

HURRICANES

_____ **24.** Which of the following best describes a hurricane?
   a. a spinning column that touches the ground
   b. a rotating tropical system with winds of 120 km/h
   c. a funnel cloud
   d. the boundary between air masses

Directed Reading A continued

Where Hurricanes Form

_____ **25.** Where do hurricanes form?
   a. over land
   b. over the Great Lakes
   c. over the Arctic Ocean
   d. over warm, tropical oceans

How Hurricanes Form

_____ **26.** What happens when a hurricane moves over cold waters?
   a. The storm gets larger.
   b. The storm's wind speed rises.
   c. The storm moves quickly.
   d. The storm loses energy.

EFFECTS OF SEVERE WEATHER

_____ **27.** Which of the following is NOT part of severe weather?
   a. drizzly rain
   b. hail
   c. flash floods
   d. high winds

_____ **28.** What is the leading cause of weather-related deaths?
   a. lightning
   b. thunder
   c. flash floods
   d. high winds

_____ **29.** What is a storm surge?
   a. strong winds and hail
   b. a rise in sea level during a storm
   c. a low-pressure area
   d. heavy rain

SEVERE-WEATHER SAFETY

_____ **30.** What should you do during a flood warning?
   a. Go to the basement.
   b. Move to higher ground.
   c. Enter the floodwaters.
   d. Do nothing.

Directed Reading A

Section: What Is Climate? (pp. 520–527)
CLIMATE VERSUS WEATHER
Write the letter of the correct answer in the space provided.

_____ **1.** What kind of conditions vary from day to day?
    a. climate
    b. weather
    c. latitude
    d. elevation

_____ **2.** What are the average weather conditions in an area over a long period of time called?
    a. latitude
    b. temperature
    c. climate
    d. wind

FACTORS THAT AFFECT CLIMATE
Solar Energy and Latitude

_____ **3.** What does latitude measure?
    a. tilt of Earth's axis
    b. heat from the sun
    c. distance from the equator
    d. air temperatures

_____ **4.** What affects the amount of direct solar energy at different latitudes?
    a. weather
    b. global circulation
    c. Earth's size
    d. the curve of Earth

_____ **5.** Why does the equator have high temperatures?
    a. The sun's rays hit it directly.
    b. It has more trees.
    c. It gets less solar heat.
    d. It has a higher latitude.

| **Directed Reading A** *continued*

Latitude and Seasons

_____ **6.** What causes seasons to happen?
   a. the equator
   b. cold weather at the poles
   c. the tilt of Earth's axis
   d. shorter days

Global Circulation and Winds

_____ **7.** What does uneven heating on Earth's surface form?
   a. the four seasons
   b. streamlike movements of water
   c. lots of precipitation
   d. areas of different air pressure

_____ **8.** Which of the following statements is NOT true about winds?
   a. Winds affect the latitude.
   b. Winds affect the amount of precipitation.
   c. Winds carry evaporated water.
   d. Winds carry solar energy.

Topography

Use the terms from the following list to complete the sentence below.

   rain-shadow effect   topography   elevation

9. The height of a landform above sea level is called _____.

10. Air that passes over a mountain cools, and precipitation may fall. This is a

   process called _____.

11. The sizes and shapes of the land-surface features of a region

   form _____.

Proximity to Large Bodies of Water

Write the letter of the correct answer in the space provided.

_____ **12.** How does a large body of water affect nearby land?
   a. The temperatures have wide ranges.
   b. There is a rain-shadow effect.
   c. Temperature changes are sudden.
   d. Sudden temperature changes are rare.

_____ **13.** What keeps California's temperatures moderate?
    a. the Great Lakes
    b. precipitation
    c. the ocean
    d. latitude

Ocean Currents

_____ **14.** What are streamlike movements of water on top of the ocean called?
    a. rain-shadow effect
    b. surface currents
    c. convection currents
    d. dew point

_____ **15.** What keeps the West Coast's temperatures cooler than inland temperatures?
    a. air masses
    b. Atlantic Ocean
    c. California current
    d. Great Lakes

CLIMATES OF THE WORLD

_____ **16.** Why do different parts of the world have such different animals?
    a. Animals adapt to only certain climates.
    b. Animals cannot travel.
    c. Climate does not affect animals.
    d. Plants cannot live in arctic areas.

Climate Zones

_____ **17.** Which of the following is NOT a major climate zone?
    a. equatorial
    b. temperate
    c. polar
    d. tropical

_____ **18.** Which of the following is the coldest climate zone?
    a. rain forest
    b. tropical
    c. polar
    d. temperate

Skills Worksheet

Directed Reading A

Section: Changes in Climate (pp. 528–533)

Write the letter of the correct answer in the space provided.

_____ 1. Why do you not notice a change in climate?
   a. because people cause it
   b. because climate changes slowly
   c. because you cannot measure climate
   d. because climate never changes

ICE AGES

_____ 2. What happens during an ice age?
   a. Earth gets warmer.
   b. Ice moves toward lower latitudes.
   c. There is less ice on Earth.
   d. The sea level gets higher.

_____ 3. What happens during a glacial period?
   a. The sea level gets higher.
   b. There is less ice on Earth.
   c. Ice sheets get bigger.
   d. Ice sheets thaw.

CAUSES OF CLIMATE CHANGE

Changes in the Earth's Orbit and Tilt

_____ 4. What did Milankovitch say were the causes of an ice age?
   a. volcanoes
   b. changes in Earth's orbit
   c. the sun's cycle
   d. the sea level rising

_____ 5. What do the three factors of Milankovitch's theory change?
   a. the movement of the continents
   b. the amount of debris from asteroids
   c. the flow of air and moisture on Earth
   d. the amount of solar radiation received by Earth

Plate Tectonics

_____ **6.** What determines how much solar radiation a continent receives?
    a. the continent's location
    b. the continent's size
    c. the continent's weather
    d. the continent's humidity

The Sun's Cycle

_____ **7.** What powers most cycles on Earth?
    a. volcanic eruptions
    b. glacial periods
    c. the sun's energy
    d. asteroids

Asteroid Impacts

_____ **8.** What is an asteroid?
    a. a mountain that may erupt
    b. dust and debris
    c. a small rocky object
    d. part of the sun's cycle

_____ **9.** What is debris?
    a. an ice age
    b. high-energy radiation
    c. a small rocky object
    d. dust and smaller rocks

_____ **10.** How might an asteroid have caused the extinction of the dinosaurs?
    a. It crashed into them.
    b. Its dust blocked the sun.
    c. Its orbit blocked the sun.
    d. It made Earth hotter.

Volcanic Eruptions

_____ **11.** How do volcanoes change the climate?
    a. Ash makes the land warmer.
    b. Fire warms the air.
    c. Dust blocks some of the sun's rays.
    d. Ice forms at high latitudes.

| Directed Reading A *continued*

Human Activity .

_____ **12.** What is a gradual increase in Earth's temperatures called?
   a. global warming
   b. ice age
   c. glacial period
   d. interglacial period

GLOBAL WARMING

_____ **13.** What is Earth's natural heating process called?
   a. interglacial period
   b. glacial period
   c. ice age
   d. greenhouse effect

The Greenhouse Effect on Earth

_____ **14.** Which of the following is a gas that causes global temperatures to rise?
   a. oxygen
   b. ice
   c. debris
   d. carbon dioxide

Consequences of Global Warming

_____ **15.** What could happen because of global warming?
   a. Icecaps could form.
   b. Deserts could be drier.
   c. Sea levels could go down.
   d. Farming in Canada could fail.

What Humans Can Do

_____ **16.** How can humans reduce pollution rates?
   a. by cutting down trees
   b. by turning up the heat
   c. by always driving automobiles
   d. by turning off the lights

Skills Worksheet

Vocabulary and Section Summary A

Water in the Air

VOCABULARY

In your own words, write a definition of the following terms in the space provided.

1. weather

_____

_____

_____

2. humidity

_____

_____

_____

3. dew point

_____

_____

_____

4. relative humidity

_____

_____

_____

5. condensation

_____

_____

_____

6. precipitation

_____

_____

_____

▌Vocabulary and Section Summary A *continued*

SECTION SUMMARY

Read the following section summary.

- The sun's energy causes water to change states and to move through the water cycle.

- The amount of water vapor in the air is called *humidity*.

- When the temperature of the air cools to the dew point, the air is saturated and liquid water droplets form.

- Clouds form as air rises and cools, which causes water droplets to form on small particles in the air.

- Precipitation is water, in any form, that falls to Earth's surface from the clouds.

Skills Worksheet

Vocabulary and Section Summary A

Fronts and Weather
VOCABULARY

In your own words, write a definition of the following terms in the space provided.

1. air mass

_____

_____

_____

2. front

_____

_____

_____

3. cyclone

_____

_____

_____

4. lightning

_____

_____

_____

5. thunder

_____

_____

_____

6. tornado

_____

_____

_____

| Vocabulary and Section Summary A *continued*

SECTION SUMMARY

Read the following section summary.

- Thunderstorms are weather systems that produce strong winds, heavy rain, lightning, and thunder.

- Lightning is a large electric discharge that occurs between two oppositely charged surfaces. Lightning releases a great deal of energy and can be very dangerous.

- Tornadoes are small, rapidly rotating columns of air that touch the ground and can cause severe damage.

- A hurricane is a large, rotating tropical weather system that has high wind speeds.

- In the event of severe weather, it is important to stay safe. Listen to your local TV or radio stations for updates, and remain indoors and away from windows.

Vocabulary and Section Summary A

What Is Climate?

VOCABULARY

In your own words, write a definition of the following terms in the space provided.

1. climate

_____

_____

2. latitude

_____

_____

3. elevation

_____

_____

4. surface current

_____

_____

SECTION SUMMARY

Read the following section summary.

- Climate is the average weather conditions in an area over a long period of time.
- The curve of Earth's surface affects the amount of direct solar energy that reaches the ground at different latitudes.
- The tilt of Earth on its axis causes the seasons.
- Winds affect the climate of an area by transferring solar energy and by carrying moisture.
- Topography influences an area's climate by affecting both the transfer of energy and the rate of precipitation.
- Large bodies of water and ocean currents influence the climate of an area by affecting the temperature of the air.
- The three main climate zones of the world are the tropical zone, the temperate zone, and the polar zone.

Skills Worksheet

Vocabulary and Section Summary A

Changes in Climate
VOCABULARY
In your own words, write a definition of the following terms in the space provided.

1. ice age

_____

_____

_____

2. global warming

_____

_____

_____

3. greenhouse effect

_____

_____

_____

SECTION SUMMARY

Read the following section summary.

- Earth's climate has changed repeatedly throughout geologic time.
- The Milankovitch theory states that Earth's climate changes as Earth's orbit and the tilt of Earth's axis change.
- Climate changes may also be affected by continental drift, volcanic eruptions, asteroid impacts, solar activity, and human activity.
- Excess CO_2 is believed to contribute to global warming.

Skills Worksheet

Directed Reading A

Section: Everything Is Connected (pp. 550–553)
STUDYING THE WEB OF LIFE

Write the letter of the correct answer in the space provided.

_____ **1.** What is the study of interactions of organisms and their environment called?
- **a.** population
- **b.** ecology
- **c.** specialization
- **d.** organism

The Two Parts of an Environment

Use the terms from the following list to complete the sentences below.

abiotic   biotic

2. Animals and all living things are _____ parts of the environment.

3. Water, rocks, and other nonliving things are

_____ parts of the environment.

Organization in the Environment

Write the letter of the correct answer in the space provided.

_____ **4.** How many levels can the environment be arranged into?
- **a.** five
- **b.** six
- **c.** four
- **d.** seven

Populations

_____ **5.** What are individuals of the same species that live together called?
- **a.** abiotic
- **b.** decomposers
- **c.** a population
- **d.** a biome

| Directed Reading A *continued*

Communities

_____ **6.** What do all the populations of organisms that live and interact in an area make up?
   a. a city
   b. a population
   c. a community
   d. an individual

Ecosystems

_____ **7.** What is a community of organisms and their abiotic environment called?
   a. an individual
   b. an ecosystem
   c. a population
   d. a community

Biomes

_____ **8.** What is an area where the climate determines the kinds of plants and animals that live there called?
   a. a biosphere
   b. a biome
   c. a community
   d. a population

_____ **9.** What would happen to some desert plants if they were placed in a biome that receives a lot of snow?
   a. They would grow larger.
   b. They would change color.
   c. They would not survive.
   d. They would spread.

The Biosphere

_____ **10.** What is the biosphere?
   a. a group of animals that live together
   b. different species of animals in a community
   c. the part of Earth where life exists
   d. the rivers of a salt marsh

_____ **11.** Where does the biosphere exist?
   a. from deepest ocean to high in the air
   b. only from the ocean to Earth's surface
   c. only on top of Earth's surface
   d. only from Earth's surface to the tops of mountains

Skills Worksheet

Directed Reading A

Section: Living Things Need Energy (pp. 554–559)

Write the letter of the correct answer in the space provided.

_____ 1. What do all living things need to survive?
   a. plants
   b. animals
   c. organisms
   d. energy

THE ENERGY CONNECTION

_____ 2. What are the three groups of living things?
   a. abiotic, biotic, neutral
   b. producers, consumers, decomposers
   c. energy, no energy, variable energy
   d. grasslands, prairies, water

Producers

Use the terms from the following list to complete the sentences below.

   producers   photosynthesis

3. Organisms that use sunlight to make food

   are called _____.

4. The process of making sunlight into food is called _____.

Consumers

Match the correct description with the correct term. Write the letter in the space provided.

_____ 5. a consumer that eats only plants   **a.** herbivore

_____ 6. a consumer that eats only animals   **b.** scavenger

                                                       c. carnivore
_____ 7. a consumer that eats both plants and animals
                                                       d. omnivore

_____ 8. an omnivore that eats dead things

| **Directed Reading A** *continued*

Decomposers

Write the letter of the correct answer in the space provided.

_____ **9.** Which of the following are organisms that get energy by breaking
down dead organisms?
   a. materials
   b. carbon dioxide
   c. decomposers
   d. water

_____ **10.** What do decomposers produce?
   a. water and carbon dioxide
   b. an ecosystem
   c. food from sunlight
   d. consumers

Food Chains

_____ **11.** What kind of diagram shows how energy is transferred from one
organism to another?
   a. producer
   b. ecology
   c. consumer
   d. food chain

_____ **12.** What kind of organisms eat secondary consumers?
   a. tertiary consumers
   b. primary consumers
   c. prey
   d. decomposers

Food Webs

_____ **13.** What does a food web show?
   a. the process of photosynthesis
   b. the feeding relationships between organisms
   c. the limiting factor of organisms
   d. an organism's environment

_____ **14.** How does energy move in a food web?
   a. not at all
   b. back and forth
   c. in two directions
   d. in one direction

| Directed Reading A *continued*

Energy Pyramids

_____ **15.** What triangle-shaped diagram shows how energy is lost?
   a. food web
   b. food chain
   c. energy pyramid
   d. community

_____ **16.** Which of the following best describes the shape of an energy pyramid?
   a. a large base and a large top
   b. a small base and a small top
   c. a large base and a small top
   d. a small base and a large top

WOLVES AND THE ENERGY PYRAMID

_____ **17.** In some parts of the United States, what happened to the elk
   population when the wolf population became smaller?
   a. The elk ate more animals.
   b. The elk left the wilderness.
   c. The elk overgrazed the grass.
   d. The elk died out.

Balance in Ecosystems

_____ **18.** Why were gray wolves brought back to Yellowstone National Park?
   a. to help the elk
   b. to kill the old and sick elk
   c. to keep the grass from taking over
   d. to eat the cows and sheep

_____ **19.** What happened to the elk population in Yellowstone when the wolves
   came back?
   a. It stayed the same.
   b. It got larger.
   c. It got smaller, then larger.
   d. It got smaller.

Directed Reading A

Section: Types of Interactions (pp. 560–565)
INTERACTIONS WITH THE ENVIRONMENT
Write the letter of the correct answer in the space provided.

_____ 1. What parts of the environment can control the size of a population?
   a. biotic and abiotic factors
   b. fish and frogs
   c. small ponds
   d. seaweed and fish

Limiting Factors

_____ 2. Why can't populations grow without stopping?
   a. Food and water are limited.
   b. Food and water are everywhere.
   c. Animals are not strong.
   d. Air is limited.

_____ 3. What is an example of a limiting factor?
   a. a very small population
   b. not enough food for a population
   c. too much food for a population
   d. too much water for a population

Carrying Capacity

_____ 4. What is the largest population an environment can support called?
   a. the limiting factor
   b. the interaction factor
   c. the population capacity
   d. the carrying capacity

_____ 5. What happens when a population grows larger than its carrying capacity?
   a. The population increases more.
   b. Individuals die.
   c. Individuals grow larger.
   d. There is less rainfall.

INTERACTIONS BETWEEN ORGANISMS

_____ **6.** Which one of the following relationships is NOT a way that species affect each other?
   a. competitive relationships
   b. capacity relationships
   c. predator and prey relationships
   d. symbiotic relationships

COMPETITION

_____ **7.** What is it called when individuals or populations try to use the same resource?
   a. coevolution
   b. competition
   c. capacity
   d. a community

_____ **8.** What kind of competition takes place when trees compete for the same space?
   a. competition between populations
   b. competition between communities
   c. competition within populations
   d. competition within communities

PREDATORS AND PREY

Use the terms from the following list to complete the sentences below.

   predator   prey

9. When a bird eats a worm, the worm is the _____.

10. When a bird eats a worm, the bird is the _____.

| **Directed Reading A** *continued* |

Predator Adaptations

Write the letter of the correct answer in the space provided.

_____ **11.** What must all predators do to survive?
   a. chase wolves
   b. catch their prey
   c. wash daily
   d. make nests

_____ **12.** What do cheetahs do to catch prey?
   a. hide quietly
   b. spin webs
   c. run very quickly
   d. catch spiders

_____ **13.** How do goldenrod spiders catch their prey?
   a. blend in with the flowers
   b. chase insects
   c. run very quickly
   d. sting insects

Prey Adaptations

_____ **14.** What do small fishes do to protect themselves from predators?
   a. hide in weeds
   b. attack the predators
   c. swim in schools
   d. have bright colors

_____ **15.** How do fire salamanders protect themselves from predators?
   a. spray poison
   b. hide in flowers
   c. run fast
   d. run into burrows

Camouflage

_____ **16.** What is blending in with the background called?
   a. avoidance
   b. camouflage
   c. community
   d. color

_____ **17.** How do walking stick insects camouflage themselves?
   a. They build stick houses.
   b. They look like twigs or sticks.
   c. They look like stones.
   d. They freeze against a shrub.

Directed Reading A *continued*

Defensive Chemicals

_____ **18.** How do the skunk and the bombardier beetle save themselves?
    a. They camouflage and hide.
    b. They are able to run quickly.
    c. They spray irritating chemicals.
    d. They freeze against shrubs.

Warning Coloration

_____ **19.** How do some animals advertise that they have a chemical defense?
    a. They make a lot of noise.
    b. They try to look innocent.
    c. They stay near other similar animals.
    d. They have warning colors such as bright red.

SYMBIOSIS

_____ **20.** What is symbiosis?
    a. a distant association between organisms
    b. a close, long-term association between two species
    c. a close, short-term association between two species
    d. a chemical association between organisms

Mutualism

_____ **21.** What is the symbiotic relationship in which both organisms help each other?
    a. mutualism
    b. commensalism
    c. parasitism
    d. camouflage

Commensalism

_____ **22.** In what symbiotic relationship does one living thing benefit while the other is not affected?
    a. mutualism
    b. camouflage
    c. ecosystem
    d. commensalism

_____ **23.** How do sharks and remoras show commensalism?
    a. Remoras ride on sharks.
    b. Remoras eat sharks.
    c. Sharks eat remoras.
    d. Remoras save sharks.

| Directed Reading A *continued*

Parasitism

Use the terms from the following list to complete the sentences below.

    parasitism   parasite   host

24. A symbiotic relationship in which one organism benefits and the other

    is harmed is called _____.

25. A parasite gets nourishment from its _____.

26. The host is weakened by the _____.

Name _____ Class _____ Date _____

Vocabulary and Section Summary A

Everything Is Connected
VOCABULARY

In your own words, write a definition of the following terms in the space provided.

1. ecology

_____

_____

2. abiotic

_____

_____

3. biotic

_____

_____

4. population

_____

_____

5. community

_____

_____

6. ecosystem

_____

_____

7. biome

_____

_____

8. biosphere

_____

_____

SECTION SUMMARY

Read the following section summary.

- Organisms in an ecosystem depend on other organisms and on abiotic factors for their survival.

- Energy and other resources flow between organisms and their environment.

- Biotic factors are the interactions between organisms in an area, such as competition.

- Abiotic factors include all of the nonliving things in an area, such as water and light.

Name _____ Class _____ Date _____

Vocabulary and Section Summary A

Living Things Need Energy
VOCABULARY

In your own words, write a definition of the following terms in the space provided.

1. herbivore

_____

_____

2. carnivore

_____

_____

3. omnivore

_____

_____

4. food chain

_____

_____

5. food web

_____

_____

6. energy pyramid

_____

_____

_____

Vocabulary and Section Summary A *continued*

SECTION SUMMARY

Read the following section summary.

- Producers that use photosynthesis transfer the energy in sunlight into chemical energy.

- Consumers eat producers and other organisms to obtain energy and nutrients.

- Decomposers break down all of the materials in dead organisms to obtain energy and nutrients.

- Food chains represent how energy and nutrients are transferred from one organism to another.

- Energy pyramids show how energy is lost at each level of the food chain.

- All organisms have important roles in a food web.

Skills Worksheet

Vocabulary and Section Summary A

Types of Interactions
VOCABULARY

In your own words, write a definition of the following terms in the space provided.

1. carrying capacity

_____

_____

_____

2. mutualism

_____

_____

3. prey

_____

_____

4. predator

_____

_____

5. symbiosis

_____

_____

_____

6. commensalism

_____

_____

_____

7. parasitism

_____

_____

_____

❚ Vocabulary and Section Summary A *continued*

SECTION SUMMARY

Read the following section summary.

- Limiting factors refer to the factors in the environment that keep a population from growing without limits.

- Competition happens within populations and between populations.

- A predator is an organism that kills an organism to eat all or part of that organism. The organism that is killed and eaten is called *prey*.

- Prey have features such as camouflage, chemical defenses, and warning coloration that protect them from predators.

- Mutualism, commensalism, and parasitism are the three kinds of symbiotic relationships that occur between organisms.

Directed Reading A

Section: Studying the Environment (pp. 580–583)

Write the letter of the correct answer in the space provided.

_____ **1.** What are the nonliving factors of the environment called?
  a. deposits
  b. abiotic
  c. organic
  d. ecosystems

_____ **2.** What are the living parts of the environment called?
  a. deposits
  b. abiotic
  c. organisms
  d. biomes

HOW BIOMES DIFFER FROM ECOSYSTEMS

_____ **3.** What are biomes made up of?
  a. deposits
  b. climate
  c. ecosystems
  d. communities

_____ **4.** What defines an ecosystem?
  a. organisms and abiotic factors
  b. rainfall and temperature
  c. abiotic factors and temperature
  d. land and water

_____ **5.** What determines which plants will be found in a biome?
  a. the climate
  b. the wind
  c. the animals
  d. the soil

_____ **6.** What are two factors that make up climate?
  a. plants and wind
  b. rain and temperature
  c. rain and animals
  d. soil and plants

| Directed Reading A *continued*

ABIOTIC FACTORS OF AN ENVIRONMENT

_____ **7.** What do producers need for photosynthesis?
  a. sunlight
  b. rainfall
  c. organisms
  d. temperature

_____ **8.** At what kind of temperatures do most organisms function best?
  a. cold   **c.** cool
  b. hot   **d.** warm

_____ **9.** Why are there fewer plants in a desert than in a forest?
  a. A desert has less sunlight than a forest.
  b. A forest has more sunlight than a desert.
  c. A desert has less rainfall than a forest.
  d. A desert has more rainfall than a forest.

_____ **10.** What can be added to soil to support plant growth?
  a. sunlight   **c.** temperature
  b. fertilizers   **d.** photosynthesis

ROLES OF ORGANISMS IN AN ENVIRONMENT

Match the correct description with the correct term. Write the letter in the space provided.

_____ **11.** organisms that eat other organisms   **a.** producers

_____ **12.** organisms that break down dead organisms   **b.** decomposers

_____ **13.** organisms that convert energy into food   **c.** consumers

SIMILAR CLIMATES, SIMILAR BIOMES

Write the letter of the correct answer in the space provided.

_____ **14.** What do biomes with the same kinds of plants and animals also share?
  a. similar climate
  b. similar soils
  c. similar variations
  d. similar consumers

_____ **15.** What is the biome for most of California called?
  a. ecosystem
  b. abiotic
  c. decomposer
  d. chaparral

Directed Reading A

Section: Land Biomes (pp. 584–591)

Write the letter of the correct answer in the space provided.

_____ **1.** What are the three categories of organisms in a biome?
- **a.** decomposer, producer, carnivore
- **b.** carnivore, herbivore, producer
- **c.** producer, consumer, decomposer
- **d.** herbivore, omnivore, consumer

DESERTS

_____ **2.** Which of the following describe a desert?
- **a.** humid and wet
- **b.** seasonal rains and grass
- **c.** cold and snowy
- **d.** very hot and dry

_____ **3.** How do plants adapt to the desert climate?
- **a.** They grow far apart.
- **b.** They grow only at night.
- **c.** They have roots above the ground.
- **d.** They do not have flowers.

_____ **4.** How does a fringe-toed lizard adapt to the desert?
- **a.** grows large ears
- **b.** stores water under its shell
- **c.** buries itself in loose sand
- **d.** recycles water from food

_____ **5.** How does a tortoise adapt to the desert?
- **a.** grows large ears
- **b.** stores water under its shell
- **c.** buries itself in loose sand
- **d.** recycles water from food

CHAPARRAL

_____ **6.** Which of the following does a chaparral have?
- **a.** patches of evergreen shrubs
- **b.** treeless plains
- **c.** layers of permafrost
- **d.** coniferous forests

| Directed Reading A *continued*

_____ **7.** What helps maintain the chaparral?
  a. lots of rainfall
  b. permafrost
  c. deciduous trees
  d. natural fires

_____ **8.** How do mule deer adapt to living in a chaparral?
  a. They live in the canopy.
  b. They migrate over large distances.
  c. They blend into their surroundings.
  d. They have bright colors.

GRASSLANDS

_____ **9.** What are the main plants in a grassland?
  a. conifers
  b. woody shrubs
  c. grasses
  d. evergreens

Prairies

_____ **10.** Which of the following describes a temperate grassland?
  a. soil rich in nutrients
  b. treeless plain
  c. diversity of plants
  d. very hot and dry

_____ **11.** Why are there few trees in temperate grasslands?
  a. Too much rain prevents tree growth.
  b. Fires and drought prevent tree growth.
  c. Trees cannot grow deep enough roots.
  d. The rolling hills of the grassland prevent tree growth.

_____ **12.** Which of the following is NOT found in a temperate grassland?
  a. howler monkey
  b. prairie dog
  c. bison
  d. mice

Savannas

_____ **13.** What is the climate of a savanna?
  a. rainy year-round
  b. seasonal rains
  c. always dry and hot
  d. always cold and wet

_____ **14.** Which of the following do lions in the African savanna prey on?
    a. mule deer
    b. toucan
    c. polar bear
    d. elephant

TUNDRA

_____ **15.** Which of the following describes a tundra?
    a. very hot and dry
    b. dominated by grasses
    c. treeless plain
    d. warm and rainy

_____ **16.** What is permafrost?
    a. soil that has thawed
    b. soil that is always frozen
    c. a plant in the tundra
    d. muddy soil

_____ **17.** Why do mosses and lichens in the tundra grow low to the ground?
    a. to resist the cold
    b. to grow more leaves
    c. to take in water
    d. to keep the soil fertile

_____ **18.** What role does a wolf in a tundra play?
    a. herbivore
    b. carnivore
    c. prey
    d. producer

FORESTS

_____ **19.** What kind of weather does a forest biome have?
    a. snowy
    b. rainy
    c. dry and hot
    d. cold

Coniferous Forests

_____ **20.** Where are conifer seeds produced?
    a. in cones
    b. in waxy leaves
    c. in needles
    d. in trunks

_____ **21.** What covers the needles of conifers?
    a. sap
    b. bark
    c. weeds
    d. waxy coating

_____ **22.** Why do few large plants grow beneath conifers?
    a. It is too cold.
    b. Animals eat them.
    c. Little light reaches the ground.
    d. The ground is too bare.

Temperate Deciduous Forests

_____ **23.** What does the word *deciduous* mean?
    a. to change color
    b. to fall off
    c. to decide
    d. to change seasons

_____ **24.** Why do some deciduous trees lose their leaves?
    a. to feed animals
    b. to make the soil better
    c. to save water
    d. to get more sunlight

_____ **25.** What is a forest canopy?
    a. small shrubs
    b. deciduous leaves
    c. forest floor
    d. tree tops

Tropical Rain Forests

_____ **26.** Where do most animals live in the tropical rain forest?
    a. in the canopy
    b. on the forest floor
    c. in caves
    d. in rivers and streams

_____ **27.** Why do rain-forest trees grow above-ground roots?
    a. There is too much rain.
    b. The soil is very thin.
    c. The temperatures are so warm.
    d. The soil is fertile.

Directed Reading A

Section: Marine Ecosystems (pp. 592–599)

Write the letter of the correct answer in the space provided.

_____ **1.** What are ecosystems in the ocean called?
    a. coniferous ecosystems
    b. deciduous ecosystems
    c. marine ecosystems
    d. polar ecosystems

DEPTH AND SUNLIGHT

_____ **2.** How deep does sunlight reach in the ocean?
    a. 4000 m
    b. 200 m
    c. 2000 m
    d. 400 m

_____ **3.** What are tiny organisms that float near the ocean's surface called?
    a. corals
    b. holdfasts
    c. algae
    d. phytoplankton

TEMPERATURE

_____ **4.** What happens to ocean water as the water gets deeper?
    a. The water gets warmer.
    b. The water gets colder.
    c. The water stays the same.
    d. The water gets colder and then warmer.

_____ **5.** Which ocean zone has the warmest water?
    a. deep zone
    b. thermocline
    c. surface zone
    d. middle layer

_____ **6.** Which animals have adapted to near-freezing water?
    a. barnacles
    b. animals in coral reefs
    c. fishes in polar areas
    d. whales

MAJOR ZONES IN THE OCEAN
The Intertidal Zone

_____ **7.** Where is the intertidal zone located?
   a. where the ocean meets the shore
   b. where the ocean floor starts to slope downward
   c. the ocean floor
   d. where the ocean floor drops sharply

_____ **8.** What must animals in the intertidal zone adapt to?
   a. little sunlight
   b. exposure to air
   c. cold water
   d. deep water

The Neritic Zone

_____ **9.** Where is the neritic zone located?
   a. where the ocean meets the shore
   b. where the ocean floor starts to slope downward
   c. the ocean floor
   d. where the ocean floor drops sharply

_____ **10.** Which of the following is NOT found in the neritic zone?
   a. sea turtles
   b. fishes
   c. sea urchins
   d. whales

The Oceanic Zone

_____ **11.** Where is the oceanic zone located?
   a. where the ocean meets the shore
   b. where the ocean floor starts to slope downward
   c. the ocean floor
   d. where the ocean floor drops sharply

_____ **12.** Which of the following do NOT live in the ocean zone?
   a. sharks   **c.** clams
   b. whales   **d.** fishes

The Abyssal Zone

_____ **13.** Where is the abyssal zone located?
   a. where the ocean meets the shore
   b. where the ocean floor starts to slope downward
   c. the ocean floor
   d. where the ocean floor drops sharply

_____ **14.** How do many animals in the abyssal zone get their food?
  a. from phytoplankton
  b. from material that sinks from above
  c. from animals on shore
  d. from the beaches

KINDS OF MARINE ECOSYSTEMS

_____ **15.** How does the ocean provide Earth's precipitation?
  a. through its temperatures
  b. through evaporation
  c. through wind patterns
  d. through its ecosystems

Intertidal Ecosystems

_____ **16.** Which of the following is an intertidal area?
  a. coral reef   **c.** pond
  b. rocky shore   **d.** swamp

_____ **17.** What is a holdfast?
  a. a rootlike structure
  b. a barnacle
  c. an estuary
  d. a coral reef

Estuaries

_____ **18.** Which of the following describes an estuary?
  a. a place where fresh water and salt water mix
  b. a place that is poor in nutrients
  c. a place with high salt concentrations
  d. a place with few phytoplankton

Coral Reefs

_____ **19.** What are coral reefs made up of?
  a. algae
  b. skeletons of stony corals
  c. sea stars
  d. sea urchins

_____ **20.** Where are coral reefs found?
  a. warm, shallow waters
  b. deep oceans
  c. cold, shallow waters
  d. fresh-flowing streams

Kelp Forests

_____ **21.** Which of the following primary consumers are found in a kelp forest?
  a. holdfasts
  b. sargassums
  c. stony corals
  d. sea urchins

Deep-Sea Hydrothermal Vents

_____ **22.** What do deep-sea hydrothermal vents release?
  a. calcium carbonate
  b. alae
  c. toxic chemicals
  d. phytoplankton

_____ **23.** Why are vent worms unusual?
  a. They have layers of skeleton.
  b. They have bacteria living in their bodies.
  c. They float.
  d. They attach themselves to rocks.

The Sargasso Sea

_____ **24.** Where is the Sargasso Sea found?
  a. in the middle of the Atlantic Ocean
  b. in the middle of the Pacific Ocean
  c. along the coast of California
  d. in a coral reef

_____ **25.** What are sargassums?
  a. bacteria
  b. floating rafts of algae
  c. shrimplike organisms
  d. corals

Polar Ice

_____ **26.** Why are there many phytoplankton in the polar ice ecosystem?
  a. It is very warm.
  b. There is a large amount of salt.
  c. Animals don't eat phytoplankton.
  d. The icy waters are rich in nutrients.

Directed Reading A

Section: Freshwater Ecosystems (pp. 600–603)

Write the letter of the correct answer in the space provided.

_____ 1. What is an important abiotic factor in freshwater ecosystems?
   a. how many plants there are
   b. how many fish there are
   c. how animals have adapted
   d. how quickly water moves

STREAM AND RIVER ECOSYSTEMS

_____ 2. What is water that flows from underground to Earth's surface called?
   a. swamp
   b. spring
   c. estuary
   d. tributary

_____ 3. How do tadpoles adapt to fast-moving water?
   a. They attach themselves to each other.
   b. They live in the mud.
   c. They attach themselves to rocks.
   d. They live under rocks.

Match the correct description with the correct term. Write the letter in the space provided.

_____ 4. a very strong, wide stream   **a.** tributary

_____ 5. a stream of water that joins a larger stream   **b.** river

POND AND LAKE ECOSYSTEMS

Life near Shore

_____ 6. Where is the littoral zone of a lake located?
   a. close to the edge
   b. in the middle
   c. at the bottom
   d. above the water

▌Directed Reading A *continued*

_____ **7.** How can algae grow in the littoral zone?
   a. The soil is sandy.
   b. Sunlight reaches the bottom.
   c. Oxygen levels are low.
   d. The water is fast moving.

_____ **8.** Which of the following are NOT found in the littoral zone?
   a. water lilies
   b. clams
   c. cattails
   d. sharks

Life Away from Shore

_____ **9.** What is the area of a lake or pond that is as deep as sunlight can reach?
   a. swamp
   b. deep-water zone
   c. open-water zone
   d. marsh

_____ **10.** Which of the following live in the open-water zone of a lake?
   a. snakes
   b. clams
   c. crustaceans
   d. lake trout

_____ **11.** Which of the following statements is true of the deep-water zone?
   a. It gets a lot of sunlight.
   b. It gets no sunlight.
   c. No fish live there.
   d. Cattails and rushes grow there.

_____ **12.** What kind of scavengers live in the deep-water zone?
   a. catfish
   b. whales
   c. fungi
   d. tadpoles

WETLAND ECOSYSTEMS

_____ **13.** Which of the following statements describes a wetland?
   a. Animals cannot live there.
   b. The soil contains a lot of moisture.
   c. The waters are icy.
   d. There is fast-moving water.

Marshes

_____ **14.** What is a marsh?
     a. a wetland ecosystem with many trees
     b. a wetland ecosystem with no animals
     c. a wetland ecosystem with grasses but no trees
     d. a wetland ecosystem with few grasses or reeds

Swamps

_____ **15.** Where are swamps found?
     a. in low-lying areas
     b. in the littoral zone
     c. in the deep-water zone
     d. near the ocean

HOW AN ECOSYSTEM CAN CHANGE

_____ **16.** Which of the following is NOT a way in which a lake can become a forest?
     a. The plants are eaten by many animals.
     b. Dead leaves settle to the bottom.
     c. The soils dry out.
     d. The lake is filled with sediment.

Skills Worksheet

Vocabulary and Section Summary A

Studying the Environment
VOCABULARY

None

SECTION SUMMARY
Read the following section summary.

- Biomes are made up of many connected ecosystems.

- Some abiotic factors are resources. Other abiotic factors are conditions in the environment.

- Organisms in biomes and ecosystems can be categorized as producers, consumers, or decomposers.

- Some widely-separated biomes have similar communities of plants and animals.

- Similar biomes are found in areas that have similar climates.

Vocabulary and Section Summary A

Land Biomes
VOCABULARY

In your own words, write a definition of the following terms in the space provided.

1. desert

_____

_____

2. chaparral

_____

_____

3. grassland

_____

_____

4. tundra

_____

_____

SECTION SUMMARY

Read the following section summary.

• A biome is characterized by a unique plant community. The plants, in turn, support unique animal communities.

• Plants and animals in a biome are adapted to the climate of the biome.

• Each organism in a biome can be categorized into the ecological role of a producer, a consumer, or a decomposer.

• Deserts are very dry and often very hot. Deserts support plants and animals that use little water.

• Chaparral biomes are fairly dry biomes that support dense patches of shrubs and trees. Animals in the chaparral blend into their surroundings to avoid predators.

• Tundras are cold areas that have permafrost and receive very little rainfall. Tundras support low-growing plants and few animals.

• Grasslands are areas where grasses are the main plants. Prairies have hot summers and cold winters. Savannas have wet and dry seasons.

• Three forest biomes are temperate deciduous forests, coniferous forests, and tropical rain forests.

Vocabulary and Section Summary A

Marine Ecosystems

VOCABULARY

In your own words, write a definition of the following terms in the space provided.

1. phytoplankton

_____

_____

2. estuary

_____

_____

SECTION SUMMARY

Read the following section summary.

- Abiotic factors that affect marine ecosystems are water temperature, water depth, and the amount of light that passes through the water.

- Producers convert solar energy or chemicals from the environment into food, or a form of chemical energy that can be used by organisms.

- Phytoplankton and algae form the base of most of the ocean's food chains. Bacteria that make food from only the chemicals in their environment also form the base of some food chains in the ocean.

- Four ocean zones are the intertidal zone, the neritic zone, the oceanic zone, and the abyssal zone.

- Intertidal ecosystems, coral reefs, estuaries, kelp forests, deep-sea vents, the Sargasso Sea, and polar ice are some marine ecosystems. These ecosystems have unique abiotic factors that support unique communities of organisms with different ecological roles.

Vocabulary and Section Summary A

Freshwater Ecosystems

VOCABULARY

In your own words, write a definition of the following terms in the space provided.

1. littoral zone

_____

_____

2. open-water zone

_____

_____

3. deep-water zone

_____

_____

4. wetland

_____

_____

5. marsh

_____

_____

6. swamp

_____

_____

SECTION SUMMARY

Read the following section summary.

- Each kind of freshwater ecosystem supports different communities of organisms because each ecosystem has different abiotic factors.

- Organisms can be categorized as producers, consumers, and decomposers in freshwater ecosystems.

- Changes in abiotic factors, such as an increase in sediment and a decrease in oxygen, can cause lake organisms to die. Eventually, further changes can lead to the development of a forest.